D0982438

Advance Praise for
Science, the Endless Frontier

"Rush Holt gives new life to Vannevar Bush's seminal report *Science, the Endless Frontier* by emphasizing its continued relevance to American science policy and raising issues that need reexamination—in particular, the relationship between the scientific enterprise and civil society. This is an important read for everyone who is deeply concerned about the status of science in today's discourse."

—Ernest J. Moniz, Massachusetts Institute of Technology and former U.S. Secretary of Energy

"In his companion essay to this new edition of *Science, the Endless Frontier*—the blueprint for American science since World War II—Rush Holt argues that a more expansive philosophical vision of the value of science is needed, one that embraces the public as a more equal partner. Hoorah for Holt, for having the courage to take on this important, timely issue."

—Naomi Oreskes, author of *Why Trust Science?*

"Vannevar Bush's *Science, the Endless Frontier* remains the touchstone for understanding how Americans regard basic research, why they pay for it, and what benefits they expect. These issues are more urgent than ever.

How can the United States channel science and technology for public good and social justice, not only economic growth? Rush Holt offers incisive reflections on Bush's blind spots and bold plan, making this the definitive edition for our era."

—**Angela N. H. Creager, Princeton University**

"Vannevar Bush's *Science, the Endless Frontier* had an immense influence on the development of U.S. science and technology as forces for public good. In his superb introductory essay for the reissue of this pioneering document, Rush Holt provides a deeply insightful discussion, not only of what Bush's blueprint achieved, but also what it missed and how this shortfall can be remedied."

—**John Holdren, Harvard University and former chief science advisor to President Obama**

"With the reissue of *Science, the Endless Frontier*, I am once more struck by the majesty of Vannevar Bush's vision for the role of science. His document serves as a prescient reminder as the United States faces the challenges of a new century. Adding a needed retrospective, Rush Holt's companion essay calls for a more robust conversation among STEM researchers, social scientists, and the public, and strongly resonates with my experiences in public policy."

—**Sylvester James Gates Jr., Brown University**

"Since World War II, *Science, the Endless Frontier* has been the central model for science policy in the United States. This welcome new edition combines Vannevar Bush's classic text with an urgent essay by Rush Holt on what the model can and cannot do for us in the twenty-first century. This is essential reading for anyone who cares about the future of science in America."

—**Audra J. Wolfe, author of** *Freedom's Laboratory: The Cold War Struggle for the Soul of Science*

"*Science, the Endless Frontier* is a consequential statement of principles that has been repeatedly invoked as a kind of biblical text in the political wars over federal policy for scientific research and training. Holt's first-rate introduction to this new edition captures the key elements of the report, outlines its significance and fate, and considers its implications for our time."

—**Daniel J. Kevles, author of** *The Baltimore Case: A Trial of Politics, Science, and Character*

"During his productive time in Congress and afterward, Rush Holt has been a champion of scientific research and the integration of science into policymaking. He understands what the place of science in our democratic society has been—and what it should be. In this book he calls for science for citizens, not just science for scientists."

—**U.S. Congresswoman Anna Eshoo**

"Vannevar Bush's influential text, *Science, the Endless Frontier*, laid out a roadmap for a whole new relationship between science, technology, and the federal government. In a compelling essay, Rush Holt helps us understand the broader context in which Bush prepared his now-famous report, and offers important insights into new considerations that should inform our discussions about science policy today."

—**David Kaiser, author of *Quantum Legacies: Dispatches from an Uncertain World***

"Rush Holt introduces *Science, the Endless Frontier* to a whole new audience, and shows us why we all should be reading Vannevar Bush's landmark report. Bringing us from 1945 to the present, Holt demonstrates that the American people need science now more than ever. Together, Bush and Holt give us a roadmap for U.S. science today."

—**Kei Koizumi, American Association for the Advancement of Science**

"With *Science, the Endless Frontier*, Vannevar Bush envisioned the cloistered infrastructure for U.S. science research that has prevailed for nearly eighty years. Physicist and public servant Rush Holt reintroduces this influential blueprint, reflecting on the constructive developments it inspired, as well as the troubling consequences that stem from a research culture unaccountable to the polity. Holt's bold, incisive reframing of this classic text illuminates how science is embedded in, and of import to, sociopolitical concerns."

—Alondra Nelson, president of the Social Science Research Council

"Rush Holt's essay on *Science, the Endless Frontier* celebrates science discovery and the democratization of science. He layers historical context with a deeper examination of publicly funded science, and he challenges the scientific community to be more inclusive. An important perspective from a physicist, policymaker, and advocate for science, Holt provides new insights to Vannevar Bush's text."

—Kaye G. Husbands Fealing, Georgia Institute of Technology

"This is the welcome revival of arguably the most important document in the history of U.S. science policy. Vannevar Bush distills in graceful prose his deep experience of directing research during World War II and Rush Holt's new introduction points to how science is essential for all citizens and a vibrant democracy. Now, let's get on with the work of getting science to do its part in our society."

—Robert Cook-Deegan, Arizona State University

Science,
the Endless
Frontier

Princeton University Press
Princeton and Oxford

Science, the Endless Frontier

VANNEVAR BUSH

With a companion essay by
RUSH D. HOLT

Science, the Endless Frontier. A Report to the President by Vannevar Bush, Director of the Office of Scientific Research and Development, July 1945 (United States Government Printing Office, Washington: 1945). https://www.nsf.gov/about/history/vbush1945.htm

Requests for permission to reproduce material from this work should be sent to permissions@press.princeton.edu

Published by Princeton University Press
41 William Street, Princeton, New Jersey 08540
6 Oxford Street, Woodstock, Oxfordshire OX20 1TR

press.princeton.edu

Library of Congress Cataloging-in-Publication Data
Names: Bush, Vannevar, 1890–1974, author. | Holt, Rush Dew, 1948–, author. | Princeton University Press.
Title: Science, the endless frontier / Vannevar Bush, Rush D. Holt.
Description: Princeton : Princeton University Press, [2021] | Includes bibliographical references.
Identifiers: LCCN 2020028356 (print) | LCCN 2020028357 (ebook) | ISBN 9780691186627 (hardback) | ISBN 9780691201658 (ebook)
Subjects: LCSH: Science and state—United States. | Research—United States.
Classification: LCC Q127.U6 B88 2021 (print) | LCC Q127.U6 (ebook) | DDC 338.973/06–dc23
LC record available at https://lccn.loc.gov/2020028356
LC ebook record available at https://lccn.loc.gov/2020028357

British Library Cataloging-in-Publication Data is available

Editorial: Jessica Yao, Arthur Werneck, and Maria Garcia
Production Editorial: Kathleen Cioffi
Text Design: Carmina Alvarez
Jacket Design: Amanda Weiss
Production: Jacqueline Poirier
Publicity: Sara Henning-Stout and Katie Lewis
Copyeditor: Erin Hartshorn

This book has been composed in Miller

Printed on acid-free paper. ∞

Printed in the United States of America

10 9 8 7 6 5 4 3 2 1

Contents

Science,
the Endless
Frontier

The Science Bargain

RUSH D. HOLT

The scientific enterprise has thrived in the United States. For three quarters of a century, American scientific productivity has been the envy of the world. Students from across the globe flock to American universities to take part in advances in every scientific discipline; American researchers in physical, biological, social, and behavioral sciences win international prizes and awards. Medical treatments and improvements in communication and transportation have extended and enriched lives, and products and processes emerging from public and private laboratories in the United States have revolutionized consumer, military, and social activities the world over. The fruits of scientific research in America abound, yet scientific thinking is not integrated into mainstream culture and politics.

Since the Second World War, generous financial support from the federal government to universities and research institutes for scientific research, as well as industrial investment in product development, have characterized the modern American scientific enterprise and made possible its achievements. The report *Science, the Endless Frontier* is recognized as the landmark document of this enterprise.

The author, Vannevar Bush, was the head of the White House's Office of Scientific Research and Development during the Second World War, and in that role had led the scientific effort that was widely recognized as having made Allied victory possible.[1] Large coordinated groups of scientists funded through government contracts and guided toward identified goals had produced an array of astonishing accomplishments—from transfusable blood plasma, population quantities of antibiotics like penicillin, and DDT and anti-malarials to prevent insect-borne illnesses, to radar, high-performance aircraft, proximity fuses for detonating munitions, and the atom bombs that would ultimately bring the war to a close. Bush oversaw this large and successful research and development enterprise as Roosevelt's informal science adviser and "Czar of Research."[2] As the end of the war came into view, he was one of many political and academic leaders contemplating how Americans could continue to reap the benefits of scientific research in peacetime. In late 1944, he received a request from Roosevelt to prepare a report that, he hoped, would lay the foundations of a lasting American science policy.

Written using input from dozens of prominent scientists and engineers, the resulting report was delivered to President Truman in July 1945, following President Roosevelt's death. As Bush wrote in the report, there had never before been a "national policy" to assure scientific progress. There was a deep respect in American culture for scientific empirical thinking and practical technology, and there had been government sponsorship of world-renowned scientific work from the Lewis and Clark expedition to military and civilian advances in geology, agriculture, medicine, astronomy, physics, and many other areas. But there had never been a central effort to support the broad scientific enterprise, nor a comprehensive appreciation of what science could contribute to American social and political advancement.

Science, the Endless Frontier presented an inspirational utilitarian vision of what science can bring to people. Invoking a classic theme in American culture, Bush wrote in his letter of transmittal, "The pioneer spirit is still vigorous within this nation. Science offers a largely unexplored hinterland for the pioneer who has the tools for his task. The rewards of such exploration both for the Nation and the individual are great. Scientific progress is one essential key to our security as a nation, to our better health, to more jobs, to a higher standard of living, and to our cultural progress." Welcomed by the scientific establishment, the report called on government to promote and support scientific research—especially basic research—and for a new independent national agency amply funded to oversee all research, military and civilian, biological, medical and physical, basic and applied, theoretical and experimental. It would ensure stable funding for long-term contracts and freedom of inquiry for scientists, and it would have the responsibility for the education of scientific specialists. In 1950, after years of debate, Congress would pass the National Science Foundation Act to create "a national policy for the promotion of basic research and education in the sciences," and to support through grants and contracts "basic scientific research in the mathematical, physical, medical, biological, engineering, and other sciences."

Science, the Endless Frontier is now known as the "seminal report" on American science policy,[3] hailed for leading to the "American postwar consensus" for the support of science,[4] and "one of the most influential policy documents in the nation's history."[5] Although various other individuals and organizations also influenced the emerging federal policy for science, the Bush report precipitated the debate that led to an unwritten policy that fostered decades of astounding progress of science. To consider the scientific landscape today one could well begin with an appreciative reading of the Bush report. Many of the issues raised are, in one form or another, still with us. The outcomes it shaped have both contributed to the brilliant scientific enterprise we see today and also cast shadows that our present moment has thrown into sharp relief. They deserve a closer look from today's perspective, to consider again what society needs that science could help to provide.

* * *

In *Science, the Endless Frontier* Bush laid out a strong, specific vision for the role of science in society that today receives at least partial credit for shaping several essential aspects of our modern scientific enterprise and how it functions. This vision was founded in several core ideas that informed Bush's recommendations and the apparatus that eventually emerged from the ensuing debate and legislation.

Most basically—and perhaps most famously—Bush made a powerful case that "scientific progress is essential," and without it "no amount of achievement in other directions can insure our health, prosperity, and security." Advances in science, Bush argued, could offer far-reaching benefits to individuals and to society as a whole, including "more jobs, higher wages, shorter hours, more abundant crops, more leisure for recreation, for study, for learning how to live without the deadening drudgery which has been the burden of the common man for ages past." He therefore declared "science is a proper concern of government," and that government should be organized to assure scientific progress.

Bush—an engineer by training—ultimately had in mind a particular sort of progress: technologies to meet the material needs of Americans. Bush's penchant for practical application suffuses the *Endless Frontier* report and his other writings. Much of his career involved the invention and development of electronic and mechanical devices. The same month Bush sent his report to the President, his magazine article entitled "As We May Think," which to some is even better known now than *Science, the Endless Frontier*, forecast in detail a practical device we now know as the personal computer.[6] To Bush, government support of research was essential to public welfare because, as he asserted, it would produce medical cures, computing machinery, jobs, weapons, and "better and cheaper products" like "air conditioning, rayon, and plastics."

Specifically, Bush advocated for the government to support basic research—that is, in Bush's words, a search for foundational knowledge "without thought of practical ends." He maintained that basic research fills the well "from which all practical knowledge must be drawn" and is the force that drives the entire process of research and innovation. "New products and new processes do not appear full-grown," he argued: "They are founded on new principles and new conceptions, which in turn are painstakingly developed by research in the purest realms of science." Ever since *Science, the Endless Frontier* Bush has been known as the champion of basic research, and the concept has been attributed to him of a metaphorical assembly line where the output of basic research passes through the process of applied research and then development and finally to human use. This idea has influenced much of federal funding up to the present. Although the report did not actually illustrate research and development with a one-dimensional line, Bush nonetheless clearly shared this common view. Basic research was valuable to Bush *because* it would drive the process toward tangible and practical outputs to meet all national needs.[7]

The Bush report located this research primarily in colleges and universities, to be conducted by trained scientists—the "small body of gifted men and women who understand the fundamental laws of nature." During the Second World War, Bush, with funding through the research agencies he headed, had shown that universities could produce powerfully relevant work quickly, even military weapons and systems. Placing research in universities made it possible in his post-war plans to greatly increase government funding without a proportional increase in the size of government. Bush was tolerant of national labs but had a low opinion of research directed by the military. Recommending that universities host the research was his hedge against both large government and science directed by the generals. In Bush's view, research was done better by the "voluntary collaboration of independent men." In universities, he saw a unique setting where "scientists may work in an atmosphere which is relatively free from the adverse pressure of convention, prejudice, or commercial necessity," provided with "a strong sense of solidarity and security, as well as a substantial degree of personal intellectual freedom."

Bush believed strongly that science should be guided by scientists. As presented in the report, his plan granted the scientific establishment the authority to choose what scientific projects to undertake. The new agency he proposed was to be overseen by a board of distinguished scientists, and the director was to be chosen by those representatives of the science establishment.

This was a critical part of his vision—in a sense, the defining one, but at the time Bush prepared *Science, the Endless Frontier*, his was not the only vision on the table.

Almost two years before, Senator Harley Kilgore, a first-term Democratic New Dealer from West Virginia, had introduced legislation "to create a central independent agency of government devoted exclusively to the progress and expansion of science and technology, first to win the war and later to contribute to the peace."[8] The proposed agency would coordinate all government research activities. Kilgore compared such strong government centralization and planning to public control of water and power systems, public schools, and public lands, all of which he regarded favorably. At the time, Bush had come out against the bill, believing that research should have no government "command and control" after the war.[9] Bush's aversion to Kilgore's legislative approach, which was gaining support, and his belief that research is more productive under control of scientists themselves led him to write the report.[10] Soon after the Bush report, Senator Kilgore had a full legislative plan for a national program of scientific research.[11]

The structural similarities between Kilgore's plan and Bush's were greater than the differences. Both men thought science was greatly underappreciated, underfunded, and uncoordinated in different parts of the government and scattered universities; both wanted a central funding agency that would encompass military and civilian research, would foster education and disseminate science throughout the country, and would assess and coordinate the research being done in the country's universities and institutes.[12] But the differences in their plans were philosophical more than they were administrative, and therefore fundamental. It was a debate about how science thrives and what its relationship to society should be. Bush's plan was predicated on autonomy for scientists, aiming to provide them with independent leadership drawn from prominent universities and complete freedom of inquiry in choosing and pursuing their research queries. Kilgore, at heart a populist, advocated a system that would be more accountable to the larger society, with an agency governed by a committee consisting of ordinary citizens, labor leaders, and educators as well as scientists, and a director, not necessarily a scientist, appointed by the president. He wanted research to address directly the nation's social and economic needs, and he wanted funding deliberately distributed around the country. Patents from the research would belong to the public. In short, Kilgore wanted an agency closer to the political processes so that it could be guided by people's perceived needs, while Bush wanted an agency more expert-driven and insulated from the kind of public control that liberal political circles advocated. This difference is illustrated by their different attitudes to social science. Although Bush envisioned a comprehensive agency overseeing all science, he excluded social and behavioral sciences, believing the social sciences were in practice too closely associated with politics and government. Kilgore took the opposite position.[13]

The debate continued in Congress and in the science community for years. Reverberations of the debate persisted for decades in policy debates in the United States and in ideological debates of the Cold War.[14] However, by 1950 the National Science Foundation Act established that the agency would be overseen by a board of distinguished scientists. The legislative outcome did not provide exactly for the arrangement Bush had called for. But today, the scientific community, usually within individual disciplines and often the researchers themselves, largely make decisions of planning, selection, and evaluation of research throughout universities and the government.

Following the report, a science policy emerged, though unwritten and imprecise. The federal government provided increasing funding to the National Science Foundation and other agencies. Federal agencies and professional associations tracked and touted scientific progress. Programs were developed for governing the thriving scientific enterprise. Congress created new committees. Science policy grew into a field of academic study. By the 1960s federal support for research and development had grown by more than 20 fold[15] from 1940, comprising nearly 2 percent of total economic activity.

The investment by the public and private sectors in scientific research and science education since 1945—cumulatively in the trillions of dollars[16]—has returned large benefits in medical cures and extended lives, increased economic productivity, eagerly received consumer and professional conveniences, and military power—just as Bush projected. Federal funding has supported an astonishing explosion in our knowledge about every aspect of our universe, world, and human physiology, society, and psychology. To choose a few examples out of hundreds of thousands: Human traits, at first thought to be a straightforward expression of inherited parental DNA, have been shown to be influenced epigenetically by parental environment. Astrophysicists have observed colliding neutron stars creating the heavy elements we find on Earth. Macroscopic quantum entanglements have produced simultaneous changes in systems widely separated from each other. Gun violence has been characterized on epidemiological and psychological grounds. Geoscientists have explained how movements of tectonic plates carry biological carbon compounds and organisms deep into the Earth. Irrational economic behavior and human implicit biases are recognized, categorized, and predicted. Individual brain cell activity can be observed instantaneously as creatures think and process stimuli. Emission and removal of carbon in the Earth's atmosphere is understood in detail. The public has had a vague notion of the creativity, beauty, and power of these advances and has wanted them to continue.

Yet the commitment to increasing federal funding—the large piece of the Bush implicit policy bargain—could not be sustained. From 1968 to 1971, as costs of the war in southeast Asia soared, federal research spending fell 10 percent (in inflation-adjusted dollars) with spending designated for basic research falling even more (by 18 percent from 1967 to 1975). Today federal spending for research and development (R&D) is less than 40 percent of what it was in the 1960s, as a percentage of the gross domestic product. Spending on R&D in industrial corporations has grown, keeping the overall spending at approximately 2.5 percent of gross domestic product (GDP) since 1968.[17] But as Bush had pointed out, corporate funding is almost entirely short-term and developmental, supporting commercially predictable outputs benefiting the investing corporation, rather than innovative research that anyone could build on. Compared with many other countries, the US investment in scientific research, once an international benchmark, has not kept pace. At least seven other countries surpass the United States in public R&D funding as a percentage of domestic economy. Scientists, seeing federal funding, though large, as far less than optimal, continually lobby for more funding. There have been occasional spurts of support— for example, in connection with the space race and the Apollo program in the 1960s, the biomedical boom and doubling of the budget of the National Institutes of Health two decades ago, and the 2009 economic stimulus. It remains to be seen whether recent proposals for increased R&D budgets, such as the conveniently named Endless Frontier bill, will be realized in the federal appropriations process. In any case, although the federal funding of R&D is much less than could be spent productively, the post–Second World War commitment to funding of science changed the landscape for science permanently.

* * *

Nevertheless, there is reason to ask: Is science providing all it should, and are citizens receiving what they need from science? Bush wrote that scientific progress was essential in the war against disease and could improve public health—yet a thriving scientific enterprise has not prevented millions of people from putting their children at calculable risk by failing to get vaccinations. Nor has the scientific progress been enough to prepare the United States to deal with a major virus pandemic in 2020. And it has not resulted in the United States undertaking the corrective measures required to stem costly climate change.

Evidently, our scientific enterprise is failing to give citizens some important things they need. These have not been failures of research—in immunology, virology, epidemiology, oceanography, or atmospheric science. Rather, they have been failures in the relationship between science and the public—something that the Bush report and subsequent debate largely overlooked.

From the modern perspective, in this regard Bush turns out to have been somewhat shortsighted. In the belief that scientific progress ultimately relies on the freedom of scientists to pursue basic research without thought of practical ends, he promoted a system that—while helping research to flourish—has also had the effect of distancing science from the public, and vice versa. His goal was to ensure not only rational, stable funding for scientists, but also the freedom to do their chosen work, unencumbered by societal direction or governmental planning.[18] While his competitor Kilgore had proposed an arrangement for all science funded by the government to be "a true servant of the people,"[19] what has resulted can be seen to be more a servant of the scientists—a system to fund work that scientists themselves choose to do.

* * *

Indeed, many scientists are convinced that they would lose scientific creativity and effectiveness if they focused where the public might ask, rather than where their trained curiosity and established research avenues take them. In my career as a research scientist and as a policymaker serving in Congress for sixteen years I have observed that scientists fiercely guard their prerogative to choose the research agenda. Though they will make some allowances in order to secure funds, they generally believe that the fruits of their independent investigations will accrue best to the public without explicit public guidance. Research grants, usually awarded through scientific review, tend to be concentrated along elite, established patterns. The scientific community, as they have sought to avoid constraints that might come from government planning, have asserted independence in a way that results in the public regarding science as beyond their ability to judge or control, or sometimes even to understand—much less participate in.

Bush called for access to higher education and scientific training to be established through a scholarship program with the goal of "encouraging and enabling a larger number of young men and women of ability to take up science as a career." This idea of select, trained researchers as the embodiment of science is reflected in the current practice of science and science education, as well as in public attitudes toward science. Researchers and their funders typically see their job as exclusively to do research. Even now most programs in science education still focus primarily on identifying and training future professional scientists and engineers, commonly called "filling the pipeline."[20] When legislators speak of our science teaching, they commonly allude to Americans' comparative disadvantage to rivals in the number of scientists and engineers.

The result is that the public sees science not as a comprehensible approach toward understanding available to them, but rather as what researchers do in their inaccessible labs. They see scientists as people who have mastered complicated ideas and instruments unfathomable to nonscientists. Products, cures, and other material benefits may emerge from research, after several unseen steps, and the receiving public has little understanding of how they came about. They see little place for themselves in science, and although they welcome practical products that emerge from the scientific enterprise, they see little place for science and scientific thinking in their lives. This presents a problem when many of the world's most urgent challenges, for example, pandemics or climate change, desperately require the public to engage with science and also to build an understanding and trust of scientists and scientific work. If members of the public think science is not intended for them, they turn away. They may not recognize situations where scientific input can help them fulfill their civic functions. They may not ask for verification of information given to them.

At the root of the issue is a limited view, traceable in part to Bush's report, of what science is and how it contributes to society. In *Science, the Endless Frontier*, Bush identified science with research and development, and its benefit to society with its more or less tangible outputs: technologies, medicine, products. But there is more to science than research, with its specialization and sometimes esoteric techniques, and the tangible outputs are only part of what the public should obtain from the science bargain and only part of what they should think of when they think of science. In its essence, science is a way of asking questions that leads to the most reliable knowledge about how things are. This is its most essential contribution.

Months before *Science, the Endless Frontier* was issued, the American Association for the Advancement of Science's Committee on Science and Society offered a more expansive, philosophical view of the place of science in a republic. The committee began, as would the Bush report a few months later, with the assertion that science is essential. The committee took a different tack in explaining how science is essential. They represented the idea, coming out of the Enlightenment, that science can have great social and political usefulness as a way of thinking. The committee declared that "a research policy is . . . as necessary to our survival as a foreign policy and a defense policy," and that the place of science in a modern government is "no less than [that of] the law and the courts." From the observation that the "discrepancy between our advanced technology and established practices and organization is one of the major threats to our free, democratic social order," the Committee continued that scientists must "build public interest into their research." The Committee on Science and Society summarized, "A policy for research and an understanding of the relation of science to society is more than a question of gadgets and even technology." It is "not a program of planning and control" but "a declaration of purpose . . . to use the instruments of critical thinking and trained organized intelligence" to strengthen "our free, democratic social order."[21]

This view did not dominate in the policy debate emanating from *Science, the Endless Frontier*. In the years after his report Bush challenged the idea of a definable and teachable scientific method that could be used outside of the laboratory by nonexperts. He recommended that, rather than trying to prepare the nonscientific public to apply critical scientific thinking to public problems, the country should put more highly talented scientists to work in research that ultimately would benefit the public.[22] While the array of material benefits from scientific research has been a rich return to society, this bargain is limited. Beyond passive public receipt of products and cures from science, we need a much greater engagement of the public with science—in both directions. We need more public guidance and oversight of science, and we need more public use of scientific thinking in public affairs. This requires action on the parts of both the public and the scientific community.

The stakes of this issue have been amply, and tragically, illustrated by the coronavirus pandemic surging in 2020. For decades previously, scientific experts had been writing alarming articles about the devastation likely from emerging diseases and about the need for public health preparations. Yet America was not prepared. In both the long term and the short, policymakers failed to provide for adequate testing, medical equipment, and trained personnel. The public had failed to demand them. The public and their policymakers generally were slow to comprehend and adopt recommended measures like social distancing or the wearing of masks. For their part, scientists had also failed to pursue some avenues of research needed to combat the virus. Given the freedom to set their own priorities, virologists undertook molecular analysis of the structure and replication of viruses, yet some of the most relevant research about viral transmission—research more connected to society at large—remained neglected. Neither the virus researchers nor the policymakers fully integrated social sciences into their thinking about possible outbreaks and consequences. Policymakers and the public had not insisted on having a part in setting the research priorities. There are many lessons to be drawn for a national science policy, the greatest of which is that a well-funded and trained cadre of scientific researchers is no substitute for an informed, engaged public. This lesson extends far beyond the example of the pandemic.

Funded research has thrived, but public evidence-based thinking and the public ability to act with comprehension on scientific evidence has not grown alongside the growth of research. Providing funding for research and allowing scientists to spend the money to pursue the activities they choose is considerably easier than applying research findings and scientific decision making to public issues. Most legislation and policy decisions have components that can be improved by incorporation of scientifically verified information. Frequently, that information is not incorporated well, or at all. It is not enough to have some scientifically trained staff in policy-making and regulatory agencies. In matters of transportation, communication, migration, agriculture, public health, and other areas, policy makers sometimes take actions not supported by generally accepted scientific understanding. It is sometimes willful neglect, although more often it is failure to recognize the relevant information. In Congress more scientific problems are embedded, yet unrecognized, in the work of so-called non-science committees than is explicit in the work of the science committees. Examples abound. Legislation on voting procedures was not regarded until recently as something requiring input from computer scientists. Economics is frequently not regarded as an empirical science for which legislators must demand evidence. It is an interesting exercise to look at the schedule of congressional hearings for a week, consider whether the subject of each could be illuminated by scientific findings, and then see whether the witness list is likely to provide those scientific findings.

It is not only the policy experts who need science. Citizens live in a different world than in

1945. Our world is more technological. Public issues of environmental protection, energy generation, healthcare delivery, and even democratic voting procedures and equipment should be illuminated by scientific research. So too should issues of pandemics, climate change, and vaccination of children. Citizens must evaluate environmental quality, hazards, and new technologies. Information is plentiful but fragmented and easily distorted. Distanced from science, citizens struggle to distinguish among validated evidence and raw opinion, wishful thinking, or deliberate misinformation.[23] Significant numbers of the public excuse or ignore glaring omissions of scientific evidence from policy making and policy statements. As we have seen with the viral pandemic, the success of any policy depends on the engagement of the public. They must be able to judge the credibility of scientific claims by asking if the claims are based on openly vetted evidence.

Moreover, they must ensure that science is included in public decisions.

If scientists are detached, however, and the public has a sense that evidence belongs only to scientists with their sophisticated techniques, it is easy to see how citizens can substitute—in place of evidence—opinions, rumors, and baseless claims in our current age of rampant internet chicanery. Scientists must help counter the impression that science is for scientists alone. The National Science Foundation has made efforts to factor into grant-making considerations of societal impact and geographic distribution, but they have had only limited success. Reviving some of Kilgore's populist ideas about explicitly directing research to social needs and tying research output more to regional economic development should be considered to demonstrate that the public is the intended beneficiary of a national science policy.

Scientists must also communicate their scientific work in a way that is clear to, and usable by, policymakers and everyday citizens. This goes beyond testifying about their research before congressional committees, giving public lectures and writing popular articles, or even creating citizen science projects—though these are important activities that, we can be encouraged, more and more scientists are taking up. Above all, scientists must help the public understand that the success of science comes from evidence-based thinking, and that evidence and evidence-based thinking are available and accessible to all. To be able to distinguish evidence that has been generally verified through scientific consensus is a primary feature of what it is called science literacy.[24] It is not trivial to teach, but it is essential—and at least as important to national welfare as the material benefits Bush focused on cultivating.

The founders wrote into our Constitution language to foster "science *and* [emphasis added] the useful arts," suggesting that they saw in science something more than the useful products of invention. As the leading proponents of the ideas of Enlightenment, the founders of the United States embraced the idea of science as an essential tool for democracy. The frequency of words like "experiment" in *The Federalist Papers* (outnumbering uses of words like "democracy") indicates that they saw government itself as a process that demands citizens adopt a scientific frame of mind on public matters.[25] They understood that public answers to public questions, to be reliable, must be grounded in evidence. Anyone can demand evidence. It is a most democratic idea.

A democratic republic requires citizens willing to seek a common understanding of the actual conditions on which they can build a future by achieving the democratic balance among competing values and visions. Without such a shared understanding, citizens' values and hopes cannot rationally be expressed and realized. Science is the process of empirical observation of the actual conditions. By asking questions so they can be answered empirically, by sharing the empirical observations transparently, and by seeking verification through public and expert critique, science has proven to be the best path to reliable knowledge. Despite false steps, occasional bias, delays, and human myopia, science has provided over the long run the most reliable view of the world. Applying the empirical techniques of science, one can come to a more and more reliable understanding of the state of things—from viruses to climate to irrational economic behavior. Of course, application of science is not the entirety of good policy. Making good policy calls for ethics and values, history and tradition, aesthetics and humane considerations, and, necessarily, politics. However, scientific evidence is the starting point. Democracy is at risk when it becomes simply a contest of opinions not grounded in evidence. When one opinion is as good as another—

each asserted as strongly or even as deceptively as possible—democracy cannot survive. This is a call to science.

Vannevar Bush recognized that for science to produce in peacetime the kinds of benefits leading to military victory in 1945—munitions, medicines, and materials—it would be necessary to apply ample federal funding to vigorous research and to the training of select researchers. That happened, and the scientific enterprise has thrived. Today our society still faces challenges not unlike those Bush saw in the world of 1945, but we also face challenges that will not be overcome with ever-improving munitions, medicines, and materials. If we ask what else our society needs that science might provide, we find it is something less tangible, but no less important to our prosperity and security than the kinds of benefits Vannevar Bush foresaw. We need the broad application of the scientific essence to our public problems. This essence—

evidence-based thinking—must be applied, not just by a select few, but broadly throughout the citizenry because that is how democracy works. It must be based on a confidence that evidence-based thinking can lead to less biased understanding of the problems, grounded on the firm understanding that science has demonstrated time and again that such thinking can do just that.

Establishing a new science policy for today will not be easy and will require work by both scientists and by the citizenry. Good science and successful democracy are difficult in similar ways. The real difficulty in either lies not in technicalities, but rather in seeking and accepting evidence and its implications that challenge the positions of one's faction or tribe or challenge one's own preconceptions or wishful thinking. Scientists must accept their responsibility to act particularly in the public interest, to be responsible to the citizenry who give them the license to do their work, and to educate and to communicate their science in a way that allows non-scientists to apply their work in civic activities. Citizens in turn must show and must require a willingness to demand and to use unbiased evidence. For decades, we have had a national commitment to foster scientific progress through the expenditure of many hundreds of billions of dollars to support scientists in their research. It is high time to make a similarly large commitment to fully integrate scientific process into our public life. We must find the way to share with all people the benefits of scientific thinking. Doing so will provide the tool for democracy that never gets dull.

Science,
the Endless
Frontier

VANNEVAR BUSH

Publisher's Note

This volume reprints the main text of the report Vannevar Bush submitted to Harry S. Truman in July 1945, with a few elements rearranged in consideration of the reader's experience. Appendices containing the reports of the Medical Advisory Committee and the Committees on Science and the Public Welfare, Discovery and Development of Scientific Talent, and Publication of Scientific Information are not reprinted here, but may be found in other published editions of the report.

President Roosevelt's Letter

THE WHITE HOUSE

Washington, D. C.
November 17, 1944

DEAR DR. BUSH:

The Office of Scientific Research and Development, of which you are the Director, represents a unique experiment of team-work and cooperation in coordinating scientific research and in applying existing scientific knowledge to the solution of the technical problems paramount in war. Its work has been conducted in the utmost secrecy and carried on without public recognition of any kind; but its tangible results can be found in the communiques coming in from the battlefronts all over the world. Some day the full story of its achievements can be told.

There is, however, no reason why the lessons to be found in this experiment cannot be profitably employed in times of peace. The information, the techniques, and the research experience developed by the Office of Scientific Research and Development and by the thousands of scientists in the universities and in private industry, should be used in the days of peace ahead for the improvement of the national health, the creation of new enterprises bringing new jobs, and the betterment of the national standard of living.

It is with that objective in mind that I would like to have your recommendations on the following four major points:

First: What can be done, consistent with military security, and with the prior approval of the military authorities, to make known to the world as soon as possible the contributions which have been made during our war effort to scientific knowledge?

The diffusion of such knowledge should help us stimulate new enterprises, provide jobs for our returning servicemen and other workers, and make possible great strides for the improvement of the national well-being.

Second: With particular reference to the war of science against disease, what can be done now to organize a program for continuing in the future the work which has been done in medicine and related sciences?

The fact that the annual deaths in this country from one or two diseases alone are far in excess of the total number of lives lost by us in battle during this war should make us conscious of the duty we owe future generations.

Third: What can the Government do now and in the future to aid research activities by public and private organizations? The proper roles of public and of private research, and their interrelation, should be carefully considered.

Fourth: Can an effective program be proposed for discovering and developing scientific talent in

American youth so that the continuing future of scientific research in this country may be assured on a level comparable to what has been done during the war?

New frontiers of the mind are before us, and if they are pioneered with the same vision, boldness, and drive with which we have waged this war we can create a fuller and more fruitful employment and a fuller and more fruitful life.

I hope that, after such consultation as you may deem advisable with your associates and others, you can let me have your considered judgment on these matters as soon as convenient—reporting on each when you are ready, rather than waiting for completion of your studies in all.

Very sincerely yours,

(s) FRANKLIN D. ROOSEVELT

Dr. VANNEVAR BUSH,

Office of Scientific Research and Development,

Washington, D. C.

Letter of Transmittal

OFFICE OF SCIENTIFIC RESEARCH
AND DEVELOPMENT

1530 P Street, NW.
Washington 25, D.C.
JULY 25, 1945

DEAR MR. PRESIDENT:

In a letter dated November 17, 1944, President Roosevelt requested my recommendations on the following points:

(1) What can be done, consistent with military security, and with the prior approval of the military

authorities, to make known to the world as soon as possible the contributions which have been made during our war effort to scientific knowledge?

(2) With particular reference to the war of science against disease, what can be done now to organize a program for continuing in the future the work which has been done in medicine and related sciences?

(3) What can the Government do now and in the future to aid research activities by public and private organizations?

(4) Can an effective program be proposed for discovering and developing scientific talent in American youth so that the continuing future of scientific research in this country may be assured on a level comparable to what has been done during the war?

It is clear from President Roosevelt's letter that in speaking of science that he had in mind the natural sciences, including biology and medicine, and I have so interpreted his questions. Progress in other fields, such as the social sciences and the humanities, is likewise important; but the program for science presented in my report warrants immediate attention.

In seeking answers to President Roosevelt's questions I have had the assistance of distinguished committees specially qualified to advise in respect to these subjects. The committees have given these matters the serious attention they deserve; indeed, they have regarded this as an opportunity to participate in shaping the policy of the country with reference to scientific research. They have had many meetings and have submitted formal reports. I have been in close touch with the work of the committees and with their members throughout. I have examined all of the data they assembled and the suggestions they submitted on the points raised in President Roosevelt's letter.

Although the report which I submit herewith is my own, the facts, conclusions, and recommendations are based on the findings of the committees which have studied these questions. Since my report is necessarily brief, I am including as appendices the full reports of the committees.

A single mechanism for implementing the recommendations of the several committees is essential. In proposing such a mechanism I have departed somewhat from the specific recommendations of the committees, but I have since been assured that the plan I am proposing is fully acceptable to the committee members.

The pioneer spirit is still vigorous within this nation. Science offers a largely unexplored hinterland for the pioneer who has the tools for his task. The rewards of such exploration both for the Nation and the individual are great. Scientific progress is one essential key to our security as a nation, to our better health, to more jobs, to a higher standard of living, and to our cultural progress.

Respectfully yours,

(s) V. Bush, Director

THE PRESIDENT OF THE UNITED STATES,
The White House,
Washington, D. C.

1

Introduction

Scientific Progress Is Essential

We all know how much the new drug, penicillin, has meant to our grievously wounded men on the grim battlefronts of this war—the countless lives it has saved—the incalculable suffering which its use has prevented. Science and the great practical genius of this nation made this achievement possible.

Some of us know the vital role which radar has played in bringing the United Nations to victory over Nazi Germany and in driving the Japanese steadily back from their island bastions. Again it was painstaking scientific research over many years that made radar possible.

What we often forget are the millions of pay envelopes on a peacetime Saturday night which are filled because new products and new industries have provided jobs for countless Americans. Science made that possible, too.

In 1939 millions of people were employed in industries which did not even exist at the close of the last war—radio, air conditioning, rayon and other synthetic fibers, and plastics are examples of the products of these industries. But these things do not mark the end of progress— they are but the beginning if we make full use of our scientific resources. New manufacturing industries can be started and many older industries greatly strengthened and expanded if we continue to study nature's laws and apply new knowledge to practical purposes.

Great advances in agriculture are also based upon scientific research. Plants which are more resistant to disease and are adapted to short growing seasons, the prevention and cure of livestock diseases, the control of our insect enemies, better fertilizers, and improved agricultural practices, all stem from painstaking scientific research.

Advances in science when put to practical use mean more jobs, higher wages, shorter hours, more abundant crops, more leisure for recreation, for study, for learning how to live without the deadening drudgery which has been the burden of the common man for ages past. Advances in science will also bring higher standards of living, will lead to the prevention or cure of diseases, will promote conservation of our limited national resources, and will assure means of defense against aggression. But to achieve these objectives—to secure a high level of employment, to maintain a position of world leadership—the flow of new scientific knowledge must be both continuous and substantial.

Our population increased from 75 million to 130 million between 1900 and 1940. In some countries comparable increases have been accompanied by famine. In this country the increase has been accompanied by more abundant food supply, better living, more leisure, longer life, and better health. This is, largely, the product of three factors—the free play of initiative of a vigorous people under democracy, the heritage of great national wealth, and the advance of science and its application.

Science, by itself, provides no panacea for individual, social, and economic ills. It can be effective in the national welfare only as a member of a team, whether the conditions be peace or war. But without scientific progress no amount of achievement in other directions can insure our health, prosperity, and security as a nation in the modern world.

Science Is a Proper Concern of Government

It has been basic United States policy that Government should foster the opening of new frontiers. It opened the seas to clipper ships and furnished land for pioneers. Although these frontiers have more or less disappeared, the frontier of science remains. It is in keeping with the American tradition—one which has made the United States great—that new frontiers shall be made accessible for development by all American citizens.

Moreover, since health, well-being, and security are proper concerns of Government, scientific progress is, and must be, of vital interest to Government. Without scientific progress the national health would deteriorate; without scientific progress we could not hope for improvement in our standard of living or for an increased number of jobs for our citizens; and without scientific progress we could not have maintained our liberties against tyranny.

Government Relations to Science— Past and Future

From early days the Government has taken an active interest in scientific matters. During the nineteenth century the Coast and Geodetic

Survey, the Naval Observatory, the Department of Agriculture, and the Geological Survey were established. Through the Land Grant College acts the Government has supported research in state institutions for more than 80 years on a gradually increasing scale. Since 1900 a large number of scientific agencies have been established within the Federal Government, until in 1939 they numbered more than 40.

Much of the scientific research done by Government agencies is intermediate in character between the two types of work commonly referred to as basic and applied research. Almost all Government scientific work has ultimate practical objectives but, in many fields of broad national concern, it commonly involves long-term investigation of a fundamental nature. Generally speaking, the scientific agencies of Government are not so concerned with immediate practical objectives as are the laboratories of industry nor, on the other hand, are they as free to explore any natural phenomena without regard to possible economic applications as are the educational and private research institutions. Government scientific agencies have splendid records of achievement, but they are limited in function.

We have no national policy for science. The Government has only begun to utilize science in the nation's welfare. There is no body within the Government charged with formulating or executing a national science policy. There are no standing committees of the Congress devoted to this important subject. Science has been in the wings. It should be brought to the center of the stage—for in it lies much of our hope for the future.

There are areas of science in which the public interest is acute but which are likely to be cultivated inadequately if left without more support than will come from private sources. These areas—such as research on military problems, agriculture, housing, public health, certain medical research, and research involving expensive capital facilities beyond the capacity of private institutions—should be advanced by active Government support. To date, with the exception of the intensive war research conducted by the Office of Scientific Research and Development, such support has been meager and intermittent.

For reasons presented in this report we are entering a period when science needs and deserves increased support from public funds.

Freedom of Inquiry Must Be Preserved

The publicly and privately supported colleges, universities, and research institutes are the centers of basic research. They are the wellsprings of knowledge and understanding. As long as they are vigorous and healthy and their scientists are free to pursue the truth wherever it may lead, there will be a flow of new scientific knowledge to those who can apply it to practical problems in Government, in industry, or elsewhere.

Many of the lessons learned in the war-time application of science under Government can be profitably applied in peace. The Government is peculiarly fitted to perform certain functions, such as the coordination and support of broad programs on problems of great national importance. But we must proceed with caution in carrying over the methods which work in wartime to the very different conditions of peace. We must remove the rigid controls which we have had to impose, and recover freedom of inquiry and that healthy competitive scientific spirit so necessary for expansion of the frontiers of scientific knowledge.

Scientific progress on a broad front results from the free play of free intellects, working on subjects of their own choice, in the manner dictated by their curiosity for exploration of the unknown. Freedom of inquiry must be preserved under any plan for Government support of science in accordance with the Five Fundamentals listed on page XX.

The study of the momentous questions presented in President Roosevelt's letter has been made by able committees working diligently. This report presents conclusions and recommendations based upon the studies of these committees which appear in full as the appendices. Only in the creation of one over-all mechanism rather than several does this report depart from the specific recommendations of the committees. The members of the committees have reviewed the recommendations in regard to the single mechanism and have found this plan thoroughly acceptable.

2
The War against Disease

In War

The death rate for all diseases in the Army, including the overseas forces, has been reduced from 14.1 per thousand in the last war to 0.6 per thousand in this war.

Such ravaging diseases as yellow fever, dysentery, typhus, tetanus, pneumonia, and meningitis have been all but conquered by penicillin and the sulfa drugs, the insecticide DDT, better vaccines, and improved hygienic measures. Malaria has been controlled. There has been dramatic progress in surgery.

The striking advances in medicine during the war have been possible only because we had a large backlog of scientific data accumulated through basic research in many scientific fields in the years before the war.

In Peace

In the last 40 years life expectancy in the United States has increased from 49 to 65 years largely as a consequence of the reduction in the death rates of infants and children; in the last 20 years the death rate from the diseases of childhood has been reduced 87 percent.

Diabetes has been brought under control by insulin, pernicious anemia by liver extracts; and the once widespread deficiency diseases have been much reduced, even in the lowest income groups, by accessory food factors and improvement of diet. Notable advances have been made in the early diagnosis of cancer, and in the surgical and radiation treatment of the disease.

These results have been achieved through a great amount of basic research in medicine and the preclinical sciences, and by the dissemination of this new scientific knowledge through the physicians and medical services and public health agencies of the country. In this cooperative endeavour the pharmaceutical industry has played an important role, especially during the war. All of the medical and public health groups share credit for these achievements; they form interdependent members of a team.

Progress in combating disease depends upon an expanding body of new scientific knowledge.

Unsolved Problems

As President Roosevelt observed, the annual deaths from one or two diseases are far in excess of the total number of American lives lost in battle during this war. A large fraction of these deaths in our civilian population cut short the useful lives of our citizens. This is our present position despite the fact that in the last three decades notable progress has been made in civilian medicine. The reduction in death rate from diseases of childhood has shifted the emphasis to the middle and old age groups, particularly to the malignant diseases and the degenerative processes prominent in later life. Cardiovascular disease, including chronic disease of the kidneys, arteriosclerosis, and cerebral hemorrhage, now account for 45 percent of the deaths in the United States. Second are the infectious diseases, and third is cancer. Added to these are many maladies (for example, the common cold, arthritis, asthma and hay fever, peptic ulcer) which, though infrequently fatal, cause incalculable disability.

Another aspect of the changing emphasis is the increase of mental diseases. Approximately 7 million persons in the United States are mentally ill; more than one-third of the hospital beds are occupied by such persons, at a cost of $175 million a year. Each year 125,000 new mental cases are hospitalized.

Notwithstanding great progress in prolonging the span of life and relief of suffering, much illness remains for which adequate means of prevention and cure are not yet known. While additional physicians, hospitals, and health programs are needed, their full usefulness cannot be attained unless we enlarge our knowledge of the human organism and the nature of disease. Any extension of medical facilities must be accompanied by an expanded program of medical training and research.

Broad and Basic Studies Needed

Discoveries pertinent to medical progress have often come from remote and unexpected sources, and it is certain that this will be true in the future. It is wholly probable that progress in the treatment of cardiovascular disease, renal disease, cancer, and similar refractory diseases will be made as the result of fundamental discoveries in subjects unrelated to those diseases, and perhaps entirely unexpected by the investigator. Further progress requires that the entire front of medicine and the underlying sciences of chemistry, physics, anatomy, biochemistry, physiology, pharmacology, bacteriology, pathology, parasitology, etc., be broadly developed.

Progress in the war against disease results from discoveries in remote and unexpected fields of medicine and the underlying sciences.

Coordinated Attack on Special Problems

Penicillin reached our troops in time to save countless lives because the Government coordinated and supported the program of research and development on the drug. The development moved from the early laboratory stage to large scale production and use in a fraction of the time it would have taken without such leadership. The search for better anti-malarials, which proceeded at a moderate tempo for many years, has been accelerated enormously by Government support during the war. Other examples can be cited in which medical progress has been similarly advanced. In achieving these results, the Government has provided over-all coordination and support; it has not dictated how the work should be done within any cooperating institution.

Discovery of new therapeutic agents and methods usually results from basic studies in medicine and the underlying sciences. The development of such materials and methods to the point at which they become available to medical practitioners requires teamwork involving the medical schools, the science departments of universities, Government, and the pharmaceutical industry. Government initiative, support, and coordination can be very effective in this development phase.

Government initiative and support for the development of newly discovered therapeutic materials and methods can reduce the time required to bring the benefits to the public.

Action Is Necessary

The primary place for medical research is in the medical schools and universities. In some cases coordinated direct attack on special problems may be made by teams of investigators, supplementing similar attacks carried on by the Army, Navy, Public Health Service, and other organizations. Apart from teaching, however, the primary obligation of the medical schools and universities is to continue the traditional function of such institutions, namely, to provide the individual worker with an opportunity for free, untrammeled study of nature, in the directions and by the methods suggested by his interests, curiosity, and imagination. The history of medical science teaches clearly the supreme importance of affording the prepared mind complete freedom for the exercise of initiative. It is the special province of the medical schools and universities to foster medical research in this way—a duty which cannot be shifted to government agencies, industrial organizations, or to any other institutions.

Where clinical investigations of the human body are required, the medical schools are in a unique position, because of their close relationship to teaching hospitals, to integrate such investigations with the work of the departments of preclinical science, and to impart new knowledge to physicians in training. At the same time, the teaching hospitals are especially well qualified to carry on medical research because of their close connection with the medical schools, on which they depend for staff and supervision.

Between World War I and World War II the United States overtook all other nations in medical research and assumed a position of world leadership. To a considerable extent this progress reflected the liberal financial support from university endowment income, gifts from individuals, and foundation grants in the 20's. The growth of research departments in medical schools has been very uneven, however, and in consequence most of the important work has been done in a few large schools. This should be corrected by building up the weaker institutions, especially in regions which now have no strong medical research activities.

The traditional sources of support for medical research, largely endowment income, foundation grants, and private donations, are diminishing, and there is no immediate prospect of a change in this trend. Meanwhile, research costs have steadily risen. More elaborate and expensive equipment is required, supplies are more costly, and the wages of assistants are higher. Industry is only to a limited extent a source of funds for basic medical research.

It is clear that if we are to maintain the progress in medicine which has marked the last 25 years, the Government should extend financial support to basic medical research in the medical schools and in the universities, through grants both for research and for fellowships. The amount which can be effectively spent in the first year should not exceed 5 million dollars. After a program is under way perhaps 20 million dollars a year can be spent effectively.

3

Science and the Public Welfare

Relation to National Security

In this war it has become clear beyond all doubt that scientific research is absolutely essential to national security. The bitter and dangerous battle against the U-boat was a battle of scientific techniques—and our margin of success was dangerously small. The new eyes which radar supplied to our fighting forces quickly evoked the development of scientific countermeasures which could often blind them. This again represents the ever continuing battle of techniques. The V-1 attack on London was finally defeated by three devices developed during this war and used superbly in the field. V-2 was countered only by the capture of the launching sites.

The Secretaries of War and Navy recently stated in a joint letter to the National Academy of Sciences:

This war emphasizes three facts of supreme importance to national security: (1) Powerful new tactics of defense and offense are developed around new weapons created by scientific and engineering research; (2) the competitive time element in developing those weapons and tactics may be decisive; (3) war is increasingly total war, in which the armed services must be supplemented by active participation of every element of civilian population.

To insure continued preparedness along farsighted technical lines, the research scientists of the country must be called upon to continue in peacetime some substantial portion of those types of contribution to national security which they have made so effectively during the stress of the present war. . . .

There must be more—and more adequate—military research during peacetime. We cannot again rely on our allies to hold off the enemy while we struggle to catch up. Further, it is clear that only the Government can undertake military research; for it must be carried on in secret, much of it has no commercial value, and it is expensive. The obligation of Government to support research on military problems is inescapable.

Modern war requires the use of the most advanced scientific techniques. Many of the leaders in the development of radar are scientists who before the war had been exploring the nucleus of the atom. While there must be increased emphasis on science in the future training of officers for both the Army and Navy, such men cannot be expected to be specialists in scientific research. Therefore a professional partnership between the officers in the Services and civilian scientists is needed.

The Army and Navy should continue to carry on research and development on the improvement of current weapons. For many years the National Advisory Committee for Aeronautics has supplemented the work of the Army and Navy by conducting basic research on the problems of flight. There should now be permanent civilian activity to supplement the research work of the Services in other scientific fields so as to carry on in time of peace some part of the activities of the emergency war-time Office of Scientific Research and Development.

Military preparedness requires a permanent independent, civilian-controlled organization, having close liaison with the Army and Navy, but with funds directly from Congress and with the clear power to initiate military research which will supplement and strengthen that carried on directly under the control of the Army and Navy.

Science and Jobs

One of our hopes is that after the war there will be full employment, and that the production of goods and services will serve to raise our standard of living. We do not know yet how we shall reach that goal, but it is certain that it can be achieved only by releasing the full creative and productive energies of the American people.

Surely we will not get there by standing still, merely by making the same things we made before and selling them at the same or higher prices. We will not get ahead in international trade unless we offer new and more attractive and cheaper products.

Where will these new products come from? How will we find ways to make better products at lower cost? The answer is clear. There must be a stream of new scientific knowledge to turn the wheels of private and public enterprise. There must be plenty of men and women trained in science and technology for upon them depend both the creation of new knowledge and its application to practical purposes.

More and better scientific research is essential to the achievement of our goal of full employment.

The Importance of Basic Research

Basic research is performed without thought of practical ends. It results in general knowledge and an understanding of nature and its laws. This general knowledge provides the means of answering a large number of important practical problems, though it may not give a complete specific answer to any one of them. The function of applied research is to provide such complete answers. The scientist doing basic research may not be at all interested in the practical applications of his work, yet the further progress of industrial development would eventually stagnate if basic scientific research were long neglected.

One of the peculiarities of basic science is the variety of paths which lead to productive advances. Many of the most important discoveries have come as a result of experiments undertaken with very different purposes in mind. Statistically it is certain that important and highly useful discoveries will result from some fraction of the undertakings in basic science; but the results of any one particular investigation cannot be predicted with accuracy.

Basic research leads to new knowledge. It provides scientific capital. It creates the fund from which the practical applications of knowledge must be drawn. New products and new processes do not appear full-grown. They are founded on new principles and new conceptions, which in turn are painstakingly developed by research in the purest realms of science.

Today, it is truer than ever that basic research is the pacemaker of technological progress. In the nineteenth century, Yankee mechanical ingenuity, building largely upon the basic discoveries of European scientists, could greatly advance the technical arts. Now the situation is different.

A nation which depends upon others for its new basic scientific knowledge will be slow in its industrial progress and weak in its competitive position in world trade, regardless of its mechanical skill.

Centers of Basic Research

Publicly and privately supported colleges and universities and the endowed research institutes must furnish both the new scientific knowledge and the trained research workers. These institutions are uniquely qualified by tradition and by their special characteristics to carry on basic research. They are charged with the responsibility of conserving the knowledge accumulated by the past, imparting that knowledge to students, and contributing new knowledge of all kinds. It is chiefly in these institutions that scientists may work in an atmosphere which is relatively free from the adverse pressure of convention, prejudice, or commercial necessity. At their best they provide the scientific worker with a strong sense of solidarity and security, as well as a substantial degree of personal intellectual freedom. All of these factors are of great importance in the development of new knowledge, since much of new knowledge is certain to arouse opposition because of its tendency to challenge current beliefs or practice.

Industry is generally inhibited by preconceived goals, by its own clearly defined standards, and by the constant pressure of commercial necessity. Satisfactory progress in basic science seldom occurs under conditions prevailing in the normal industrial laboratory. There are some notable exceptions, it is true, but even in such cases it is rarely possible to match the universities in respect to the freedom which is so important to scientific discovery.

To serve effectively as the centers of basic research these institutions must be strong and healthy. They must attract our best scientists as teachers and investigators. They must offer research opportunities and sufficient compensation to enable them to compete with industry and government for the cream of scientific talent.

During the past 25 years there has been a great increase in industrial research involving the application of scientific knowledge to a multitude of practical purposes—thus providing new products, new industries, new investment opportunities, and millions of jobs. During the same period research within Government—again largely applied research—has also been greatly expanded. In the decade from 1930 to 1940 expenditures for industrial research increased from $116,000,000 to $240,000,000 and those for scientific research in Government rose from $24,000,000 to $69,000,000. During the same period expenditures for scientific research in the colleges and universities increased from \$20,000,000 to \$31,000,000, while those in the endowed research institutes declined from \$5,200,000 to \$4,500,000. These are the best estimates available. The figures have been taken from a variety of sources and arbitrary definitions have necessarily been applied, but it is believed that they may be accepted as indicating the following trends:

(a) Expenditures for scientific research by industry and Government—almost entirely applied research—have more than doubled between 1930 and 1940. Whereas in 1930 they were six times as large as the research expenditures of the colleges, universities, and research institutes, by 1940 they were nearly ten times as large.

(b) While expenditures for scientific research in the colleges and universities increased by one-half during this period, those for the endowed research institutes have slowly declined.

If the colleges, universities, and research institutes are to meet the rapidly increasing demands of industry and Government for new scientific knowledge, their basic research should be strengthened by use of public funds.

Research within the Government

Although there are some notable exceptions, most research conducted within governmental laboratories is of an applied nature. This has always been true and is likely to remain so. Hence Government, like industry, is dependent on the colleges, universities, and research institutes to expand the basic scientific frontiers and to furnish trained scientific investigators.

Research within the Government represents an important part of our total research activity and needs to be strengthened and expanded after the war. Such expansion should be directed to fields of inquiry and service which are of public importance and are not adequately carried on by private organizations.

The most important single factor in scientific and technical work is the quality of the personnel employed. The procedures currently followed within the Government for recruiting, classifying, and compensating such personnel place the Government under a severe handicap in competing with industry and the universities for first-class scientific talent. Steps should be taken to reduce that handicap.

In the Government the arrangement whereby the numerous scientific agencies form parts of larger departments has both advantages and disadvantages. but the present pattern is firmly established and there is much to be said for it. There is, however, a very real need for some measure of coordination of the common scientific activities of these agencies, both as to policies and budgets, and at present no such means exist.

A permanent Science Advisory Board should be created to consult with these scientific bureaus and to advise the executive and legislative branches of Government as to the policies and budgets of Government agencies engaged in scientific research.

This board should be composed of disinterested scientists who have no connection with the affairs of any Government agency.

Industrial Research

The simplest and most effective way in which the Government can strengthen industrial research is to support basic research and to develop scientific talent.

The benefits of basic research do not reach all industries equally or at the same speed. Some small enterprises never receive any of the benefits.

It has been suggested that the benefits might be better utilized if "research clinics" for such enterprises were to be established. Businessmen would thus be able to make more use of research than they now do. This proposal is certainly worthy of further study.

One of the most important factors affecting the amount of industrial research is the income-tax law. Government action in respect to this subject will affect the rate of technical progress in industry. Uncertainties as to the attitude of the Bureau of Internal Revenue regarding the deduction of research and development expenses are a deterrent to research expenditure. These uncertainties arise from lack of clarity of the tax law as to the proper treatment of such costs.

The Internal Revenue Code should be amended to remove present uncertainties in regard to the deductibility of research and development expenditures as current charges against net income.

Research is also affected by the patent laws. They stimulate new invention and they make it possible for new industries to be built around new devices or new processes. These industries generate new jobs and new products, all of which contribute to the welfare and the strength of the country.

Yet, uncertainties in the operation of the patent laws have impaired the ability of small industries to translate new ideas into processes and products of value to the nation. These uncertainties are, in part, attributable to the difficulties and expense incident to the operation of the patent system as it presently exists. These uncertainties are also attributable to the existence of certain abuses, which have appeared in the use of patents. The abuses should be corrected. They have led to extravagantly critical attacks which tend to discredit a basically sound system.

It is important that the patent system continue to serve the country in the manner intended by the Constitution, for it has been a vital element in the industrial vigor which has distinguished this nation.

The National Patent Planning Commission has reported on this subject. In addition, a detailed study, with recommendations concerning the extent to which modifications should be made in our patent laws is currently being made under the leadership of the Secretary of Commerce. It is recommended, therefore, that specific action with regard to the patent laws be withheld pending the submission of the report devoted exclusively to that subject.

International Exchange of Scientific Information

International exchange of scientific information is of growing importance. Increasing specialization of science will make it more important than ever that scientists in this country keep continually ahead of developments abroad. In addition a flow of scientific information constitutes one facet of general international accord which should be cultivated.

The Government can accomplish significant results in several ways: by aiding in the arrangement of international science congresses, in the official accrediting of American scientists to such gatherings, in the official reception of foreign scientists of standing in this country, in making possible a rapid flow of technical information, including translation service, and possibly in the provision of international fellowships. Private foundations and other groups partially fulfill some of these functions at present, but their scope is incomplete and inadequate.

The Government should take an active role in promoting the international flow of scientific information.

The Special Need for Federal Support

We can no longer count on ravaged Europe as a source of fundamental knowledge. In the past we have devoted much of our best efforts to the application of such knowledge which has been discovered abroad. In the future we must pay increased attention to discovering this knowledge for ourselves particularly since the scientific applications of the future will be more than ever dependent upon such basic knowledge.

New impetus must be given to research in our country. Such impetus can come promptly only from the Government. Expenditures for research in the colleges, universities, and research institutes will otherwise not be able to meet the additional demands of increased public need for research.

Further, we cannot expect industry adequately to fill the gap. Industry will fully rise to the challenge of applying new knowledge to new products. The commercial incentive can be relied upon for that. But basic research is essentially noncommercial in nature. It will not receive the attention it requires if left to industry.

For many years the Government has wisely supported research in the agricultural colleges and the benefits have been great. The time has come when such support should be extended to other fields.

In providing government support, however, we must endeavor to preserve as far as possible the private support of research both in industry and in the colleges, universities, and research institutes. These private sources should continue to carry their share of the financial burden.

The Cost of a Program

It is estimated that an adequate program for Federal support of basic research in the colleges, universities, and research institutes and for financing important applied research in the public interest, will cost about 10 million dollars at the outset and may rise to about 50 million dollars annually when fully underway at the end of perhaps 5 years.

4
Renewal of Our Scientific Talent

Nature of the Problem

The responsibility for the creation of new scientific knowledge rests on that small body of men and women who understand the fundamental laws of nature and are skilled in the techniques of scientific research. While there will always be the rare individual who will rise to the top without benefit of formal education and training, he is the exception and even he might make a more notable contribution if he had the benefit of the best education we have to offer. I cannot improve on President Conant's statement that:

". . . in every section of the entire area where the word science may properly be applied, the limiting factor is a human one. We shall have rapid or slow advance in this direction or in that depending on the number of really first-class men who are engaged in the work in question. . . . So in the last analysis, the future of science in this country will be determined by our basic educational policy."

A Note of Warning

It would be folly to set up a program under which research in the natural sciences and medicine was expanded at the cost of the social sciences, humanities, and other studies so essential to national well-being. This point has been well stated by the Moe Committee as follows:

"As citizens, as good citizens, we therefore think that we must have in mind while examining the question before us—the discovery and development of scientific talent—the needs of the whole national welfare. We could not suggest to you a program which would syphon into science and technology a disproportionately large share of the nation's highest abilities, without doing harm to the nation, nor, indeed, without crippling science. . . . Science cannot live by and unto itself alone."

. . .

"The uses to which high ability in youth can be put are various and, to a large extent, are determined by social pressures and rewards. When aided by selective devices for picking out scientifically talented youth, it is clear that large sums of money for scholarships and fellowships and monetary and other rewards in disproportionate amounts might draw into science too large a percentage of the nation's high ability, with a result highly detrimental to the nation and to science. Plans for the discovery and development of scientific talent must be related to the other needs of society for high ability. . . . There is never enough ability at high levels to satisfy all the needs of the nation; we would not seek to draw into science any more of it than science's proportionate share."

The Wartime Deficit

Among the young men and women qualified to take up scientific work, since 1940 there have been few students over 18, except some in medicine and engineering in Army and Navy programs and a few 4-F's, who have followed an integrated scientific course of studies. Neither our allies nor, so far as we know, our enemies have done anything so radical as thus to suspend almost completely their educational activities in scientific pursuits during the war period.

Two great principles have guided us in this country as we have turned our full efforts to war. First, the sound democratic principle that there

should be no favored classes or special privilege
in a time of peril, that all should be ready to sac-
rifice equally; second, the tenet that every man
should serve in the capacity in which his talents
and experience can best be applied for the pros-
ecution of the war effort. In general we have held
these principles well in balance.

In my opinion, however, we have drawn too
heavily for nonscientific purposes upon the great
natural resource which resides in our trained
young scientists and engineers. For the general
good of the country too many such men have
gone into uniform, and their talents have not
always been fully utilized. With the exception of
those men engaged in war research, all physically
fit students at graduate level have been taken
into the armed forces. Those ready for college
training in the sciences have not been permitted
to enter upon that training.

There is thus an accumulating deficit of
trained research personnel which will continue

for many years. The deficit of science and technology students who, but for the war, would have received bachelor's degrees is about 150,000. The deficit of those holding advanced degrees—that is, young scholars trained to the point where they are capable of carrying on original work—has been estimated as amounting to about 17,000 by 1955 in chemistry, engineering, geology, mathematics, physics, psychology, and the biological sciences.

With mounting demands for scientists both for teaching and for research, we will enter the post-war period with a serious deficit in our trained scientific personnel.

Improve the Quality

Confronted with these deficits, we are compelled to look to the use of our basic human resources and formulate a program which will assure their conservation and effective development. The committee advising me on scientific personnel has stated the following principle which should guide our planning:

"If we were all-knowing and all-wise we might, but we think probably not, write you a plan whereby there might be selected for training, which they otherwise would not get, those who, 20 years hence, would be scientific leaders, and we might not bother about any lesser manifestations of scientific ability. But in the present state of knowledge a plan cannot be made which will select, and assist, only those young men and women who will give the top future leadership to science. To get top leadership there must be a relatively large base of high ability selected for development and then successive skimmings of the cream of ability at successive times and at higher levels. No one can select from the bottom those who will be the leaders at the top because unmeasured and unknown factors enter into scientific, or any, leadership. There are brains and character, strength and health, happiness and spiritual vitality, interest and motivation, and no one knows what else, that must needs enter into this supra-mathematical calculus.

"We think we probably would not, even if we were all-wise and all-knowing, write you a plan whereby you would be assured of scientific leadership at one stroke. We think as we think because we are not interested in setting up an elect. We think it much the best plan, in this constitutional Republic, that opportunity be held out to all kinds and conditions of men whereby they can better themselves. This is the American way; this is the way the United States has become what it is. We think it very important that circumstances be such that there be no ceilings, other than ability itself, to intellectual ambition. We think it very important that every boy and girl shall know that, if he shows that he has what it takes, the sky is the limit. Even if it be shown subsequently that he has not what it takes to go to the top, he will go further than he would otherwise go if there had been a ceiling beyond which he always knew he could not aspire.

"By proceeding from point to point and taking stock on the way, by giving further opportunity to those who show themselves worthy of further opportunity, by giving the most opportunity to those who show themselves continually developing—this is the way we propose. This is the American way: a man works for what he gets."

Remove the Barriers

Higher education in this country is largely for those who have the means. If those who have the means coincided entirely with those persons who have the talent we should not be squandering a part of our higher education on those undeserving of it, nor neglecting great talent among those who fail to attend college for economic reasons. There are talented individuals in every segment of the population, but with few exceptions those without the means of buying higher education go without it. Here is a tremendous waste of the greatest resource of a nation—the intelligence of its citizens.

If ability, and not the circumstance of family fortune, is made to determine who shall receive higher education in science, then we shall be assured of constantly improving quality at every level of scientific activity.

The Generation in Uniform Must Not Be Lost

We have a serious deficit in scientific personnel partly because the men who would have studied science in the colleges and universities have been serving in the Armed Forces. Many had begun their studies before they went to war. Others with capacity for scientific education went to war after finishing high school. The most immediate prospect of making up some of the deficit in scientific personnel is by salvaging scientific talent from the generation in uniform. For even if we should start now to train the current crop of high school graduates, it would be 1951 before they would complete graduate studies and be prepared for effective scientific research. This fact underlines the necessity of salvaging potential scientists in uniform.

The Armed Services should comb their records for men who, prior to or during the war, have given evidence of talent for science, and make prompt arrangements, consistent with current discharge plans, for ordering those who remain in uniform as soon as militarily possible to duty at institutions here and overseas where they can continue their scientific education. Moreover, they should see that those who study overseas have the benefit of the latest scientific developments.

A Program

The country may be proud of the fact that 95 percent of boys and girls of the fifth grade age are enrolled in school, but the drop in enrollment after the fifth grade is less satisfying. For every 1,000 students in the fifth grade, 600 are lost to education before the end of high school, and all but 72 have ceased formal education before completion of college. While we are concerned primarily with methods of selecting and educating high school graduates at the college and higher levels, we cannot be complacent about the loss of potential talent which is inherent in the present situation.

Students drop out of school, college, and graduate school, or do not get that far, for a variety of reasons: they cannot afford to go on; schools and colleges providing courses equal to their capacity are not available locally; business and industry recruit many of the most promising before they have finished the training of which they are capable. These reasons apply with particular force to science: the road is long and expensive; it extends at least 6 years beyond high school; the percentage of science students who can obtain first-rate training in institutions near home is small.

Improvement in the teaching of science is imperative; for students of latent scientific ability are particularly vulnerable to high school teaching which fails to awaken interest or to provide adequate instruction. To enlarge the group of specially qualified men and women it is necessary to increase the number who go to college. This involves improved high school instruction, provision for helping individual talented students to finish high school (primarily the responsibility of the local communities), and opportunities for more capable, promising high school students to go to college. Anything short of this means serious waste of higher education and neglect of human resources.

To encourage and enable a larger number of young men and women of ability to take up science as a career, and in order gradually to reduce the deficit of trained scientific personnel, it is recommended that provision be made for a reasonable number of (a) undergraduate scholarships and graduate fellowships and (b) fellowships for advanced training and fundamental research. The details should be worked out with reference to the interests of the several States and of the universities and colleges; and care should be taken not to impair the freedom of the institutions and individuals concerned.

The program proposed by the Moe Committee in Appendix 4 would provide 24,000 undergraduate scholarships and 900 graduate fellowships and would cost about $30,000,000 annually when in full operation. Each year under this program 6,000 undergraduate scholarships would be made available to high school graduates, and 300 graduate fellowships would be offered to college graduates. Approximately the scale of allowances provided for under the educational program for returning veterans has been used in estimating the cost of this program.

The plan is, further, that all those who receive such scholarships or fellowships in science should be enrolled in a National Science Reserve and be liable to call into the service of the Government, in connection with scientific or technical work in time of war or other national emergency declared by Congress or proclaimed by the President. Thus, in addition to the general benefits to the nation by reason of the addition to its trained ranks of such a corps of scientific workers, there would be a definite benefit to the nation in having these scientific workers on call in national emergencies. The Government would be well advised to invest the money involved in this plan even if the benefits to the nation were thought of solely—which they are not—in terms of national preparedness.

5

A Problem of Scientific Reconversion

Effects of Mobilization of Science for War

We have been living on our fat. For more than 5 years many of our scientists have been fighting the war in the laboratories, in the factories and shops, and at the front. We have been directing the energies of our scientists to the development of weapons and materials and methods, on a large number of relatively narrow projects initiated and controlled by the Office of Scientific Research and Development and other Government agencies. Like troops, the scientists have been mobilized, and thrown into action to serve their country in time of emergency. But they have been diverted to a greater extent than is generally appreciated from the search for answers to the fundamental problems—from the search on which human welfare and progress depends. This is not a complaint—it is a fact. The mobilization of science behind the lines is aiding the fighting men at the front to win the war and to shorten it; and it has resulted incidentally in the accumulation of a vast amount of experience and knowledge of the application of science to particular problems, much of which can be put to use when the war is over. Fortunately, this country had the scientists—and the time—to make this contribution and thus to advance the date of victory.

Security Restrictions Should Be Lifted Promptly

Much of the information and experience acquired during the war is confined to the agencies that gathered it. Except to the extent that military security dictates otherwise, such knowledge should be spread upon the record for the benefit of the general public.

Thanks to the wise provision of the Secretary of War and the Secretary of the Navy, most of the results of war-time medical research have been published. Several hundred articles have appeared in the professional journals; many are in process of publication. The material still subject to security classification should be released as soon as possible.

It is my view that most of the remainder of the classified scientific material should be released as soon as there is ground for belief that the enemy will not be able to turn it against us in this war. Most of the information needed by industry and in education can be released without disclosing its embodiments in actual military material and devices. Basically there is no reason to believe that scientists of other countries will not in time rediscover everything we now know which is held in secrecy. A broad dissemination of scientific information upon which further advances can readily be made furnishes a sounder foundation for our national security than a policy of restriction which would impede our own progress although imposed in the hope that possible enemies would not catch up with us.

During the war it has been necessary for selected groups of scientists to work on specialized problems, with relatively little information as to what other groups were doing and had done. Working against time, the Office of Scientific Research and Development has been obliged to enforce this practice during the war, although it was realized by all concerned that it was an

emergency measure which prevented the continuous cross-fertilization so essential to fruitful
scientific effort.

Our ability to overcome possible future enemies depends upon scientific advances which
will proceed more rapidly with diffusion of
knowledge than under a policy of continued restriction of knowledge now in our possession.

Need for Coordination

In planning the release of scientific data and experience collected in connection with the war,
we must not overlook the fact that research has
gone forward under many auspices—the Army,
the Navy, the Office of Scientific Research and
Development, the National Advisory Committee
for Aeronautics, other departments and agencies
of the Government, educational institutions,
and many industrial organizations. There have
been numerous cases of independent discovery
of the same truth in different places. To permit

the release of information by one agency and to continue to restrict it elsewhere would be unfair in its effect and would tend to impair the morale and efficiency of scientists who have submerged individual interests in the controls and restrictions of war.

A part of the information now classified which should be released is possessed jointly by our allies and ourselves. Plans for release of such information should be coordinated with our allies to minimize danger of international friction which would result from sporadic uncontrolled release.

A Board to Control Release

The agency responsible for recommending the release of information from military classification should be an Army, Navy, civilian body, well grounded in science and technology. It should be competent to advise the Secretary of War and the Secretary of the Navy. It should, moreover, have sufficient recognition to secure prompt and practical decisions.

To satisfy these considerations I recommend the establishment of a Board, made up equally of scientists and military men, whose function would be to pass upon the declassification and to control the release for publication of scientific information which is now classified.

Publication Should Be Encouraged

The release of information from security regulations is but one phase of the problem. The other is to provide for preparation of the material and its publication in a form and at a price which will facilitate dissemination and use. In the case of the Office of Scientific Research and Development, arrangements have been made for the preparation of manuscripts, while the staffs under our control are still assembled and in possession of the records, as soon as the pressure for production of results for this war has begun to relax.

We should get this scientific material to scientists everywhere with great promptness, and at as low a price as is consistent with suitable format. We should also get it to the men studying overseas so that they will know what has happened in their absence.

It is recommended that measures which will encourage and facilitate the preparation and publication of reports be adopted forthwith by all agencies, governmental and private, possessing scientific information released from security control.

6
The Means to the End

New Responsibilities for Government

One lesson is clear from the reports of the several committees attached as appendices. The Federal Government should accept new responsibilities for promoting the creation of new scientific knowledge and the development of scientific talent in our youth.

The extent and nature of these new responsibilities are set forth in detail in the reports of the committees whose recommendations in this regard are fully endorsed.

In discharging these responsibilities Federal funds should be made available. We have given much thought to the question of how plans for the use of Federal funds may be arranged so that such funds will not drive out of the picture funds from local governments, foundations, and private donors. We believe that our proposals will minimize that effect, but we do not think that it can be completely avoided. We submit, however, that the nation's need for more and better scientific research is such that the risk must be accepted.

It is also clear that the effective discharge of these responsibilities will require the full attention of some over-all agency devoted to that purpose. There should be a focal point within the Government for a concerted program of assisting scientific research conducted outside of Government. Such an agency should furnish the funds needed to support basic research in the colleges and universities, should coordinate where possible research programs on matters of utmost importance to the national welfare, should formulate a national policy for the Government toward science, should sponsor the interchange of scientific information among scientists and laboratories both in this country and abroad, and should ensure that the incentives to research in industry and the universities are maintained. All of the committees advising on these matters agree on the necessity for such an agency.

The Mechanism

There are within Government departments many groups whose interests are primarily those of scientific research. Notable examples are found within the Departments of Agriculture, Commerce, Interior, and the Federal Security Agency. These groups are concerned with science as collateral and peripheral to the major problems of those Departments. These groups should remain where they are, and continue to perform their present functions, including the support of agricultural research by grants to the Land Grant Colleges and Experiment Stations, since their largest contribution lies in applying fundamental knowledge to the special problems of the Departments within which they are established.

By the same token these groups cannot be made the repository of the new and large responsibilities in science which belong to the Government and which the Government should accept. The recommendations in this report which relate to research within the Government, to the release of scientific information, to clarification of the tax laws, and to the recovery and development of our scientific talent now in uniform can be implemented by action within the existing structure of the Government. But nowhere in the Governmental structure receiving its funds from Congress is there an agency adapted to supplementing the support of basic research in the universities, both in medicine and the natural sciences; adapted to supporting research on new weapons for both Services; or adapted to administering a program of science scholarships and fellowships.

A new agency should be established, therefore, by the Congress for the purpose. Such an agency, moreover, should be an independent agency devoted to the support of scientific research and advanced scientific education alone. Industry learned many years ago that basic research cannot often be fruitfully conducted as an adjunct to or a subdivision of an operating agency or department. Operating agencies have immediate operating goals and are under constant pressure to produce in a tangible way, for that is the test of their value. None of these conditions is favorable to basic research. Research is the exploration of the unknown and is necessarily speculative. It is inhibited by conventional approaches, traditions, and standards. It cannot be satisfactorily conducted in an atmosphere where it is gauged and tested by operating or production standards. Basic scientific research should not, therefore, be placed under an operating agency whose paramount concern is anything other than research. Research will always suffer when put in competition with operations. The decision that there should be a new and independent agency was reached by each of the committees advising in these matters.

I am convinced that these new functions should be centered in one agency. Science is fundamentally a unitary thing. The number of independent agencies should be kept to a minimum. Much medical progress, for example, will come from fundamental advances in chemistry. Separation of the sciences in tight compartments, as would occur if more than one agency were involved, would retard and not advance scientific knowledge as a whole.

Five Fundamentals

There are certain basic principles which must underlie the program of Government support for scientific research and education if such support is to be effective and if it is to avoid impairing the very things we seek to foster. These principles are as follows:

(1) Whatever the extent of support may be, there must be stability of funds over a period of years so that long-range programs may be undertaken. (2) The agency to administer such funds should be composed of citizens selected only on the basis of their interest in and capacity to promote the work of the agency. They should be persons of broad interest in and understanding of the peculiarities of scientific research and education. (3) The agency should promote research through contracts or grants to organizations outside the Federal Government. It should not operate any

laboratories of its own. (4) Support of basic research in the public and private colleges, universities, and research institutes must leave the internal control of policy, personnel, and the method and scope of the research to the institutions themselves. This is of the utmost importance. (5) While assuring complete independence and freedom for the nature, scope, and methodology of research carried on in the institutions receiving public funds, and while retaining discretion in the allocation of funds among such institutions, the Foundation proposed herein must be responsible to the President and the Congress. Only through such responsibility can we maintain the proper relationship between science and other aspects of a democratic system. The usual controls of audits, reports, budgeting, and the like, should, of course, apply to the administrative and fiscal operations of the Foundation, subject, however, to such adjustments in procedure as are necessary to meet the special requirements of research.

Basic research is a long-term process—it ceases to be basic if immediate results are expected on short-term support. Methods should therefore be found which will permit the agency to make commitments of funds from current appropriations for programs of five years duration or longer. Continuity and stability of the program and its support may be expected (a) from the growing realization by the Congress of the benefits to the public from scientific research, and (b) from the conviction which will grow among those who conduct research under the auspices of the agency that good quality work will be followed by continuing support.

Military Research

As stated earlier in this report, military preparedness requires a permanent, independent, civilian-controlled organization, having close liaison with the Army and Navy, but with funds direct from Congress and the clear power to initiate military research which will supplement and strengthen that carried on directly under the control of the Army and Navy. As a temporary measure the National Academy of Sciences has established the Research Board for National Security at the request of the Secretary of War and the Secretary of the Navy. This is highly desirable in order that there may be no interruption in the relations between scientists and military men after the emergency wartime Office of Scientific Research and Development goes out of existence. The Congress is now considering legislation to provide funds for this Board by direct appropriation.

I believe that, as a permanent measure, it would be appropriate to add to the agency needed to perform the other functions recommended in this report the responsibilities for civilian-initiated and civilian-controlled military research. The function of such a civilian group would be primarily to conduct long-range scientific research on military problems—leaving to the Services research on the improvement of existing weapons.

Some research on military problems should be conducted, in time of peace as well as in war, by civilians independently of the military establishment. It is the primary responsibility of the Army and Navy to train the men, make available the weapons, and employ the strategy that will bring victory in combat. The Armed Services cannot be expected to be experts in all of the complicated fields which make it possible for a great nation to fight successfully in total war. There are certain kinds of research—such as research on the improvement of existing weapons—which can best be done within the military establishment. However, the job of long-range research involving application of the newest scientific discoveries to military needs should be the responsibility of those civilian scientists in the universities and in industry who are best trained to discharge it thoroughly and successfully. It is essential that both kinds of research go forward and that there be the closest liaison between the two groups.

Placing the civilian military research function in the proposed agency would bring it into close relationship with a broad program of basic research in both the natural sciences and medicine. A balance between military and other research could thus readily be maintained.

The establishment of the new agency, including a civilian military research group, should not be delayed by the existence of the Research Board for National Security, which is a temporary measure. Nor should the creation of the new agency be delayed by uncertainties in regard to the postwar organization of our military departments themselves. Clearly, the new agency, including a civilian military research group within it, can remain sufficiently flexible to adapt its operations to whatever may be the final organization of the military departments.

National Research Foundation

It is my judgment that the national interest in scientific research and scientific education can best be promoted by the creation of a National Research Foundation.

I. Purposes

The National Research Foundation should develop and promote a national policy for scientific research and scientific education, should support basic research in nonprofit organizations, should develop scientific talent in American youth by means of scholarships and fellowships, and should by contract and otherwise support long-range research on military matters.

II. Members

1. Responsibility to the people, through the President and Congress, should be placed in the hands of, say nine Members, who should be

persons not otherwise connected with the Government and not representative of any special interest, who should be known as National Research Foundation Members, selected by the President on the basis of their interest in and capacity to promote the purposes of the Foundation.

2. The terms of the Members should be, say, 4 years, and no Member should be eligible for immediate reappointment provided he has served a full 4-year term. It should be arranged that the Members first appointed serve terms of such length that at least two Members are appointed each succeeding year.

3. The Members should serve without compensation but should be entitled to their expenses incurred in the performance of their duties.

4. The Members should elect their own chairman annually.

5. The chief executive officer of the Foundation should be a director appointed by the Members. Subject to the direction and supervision of

the Foundation Members (acting as a board), the director should discharge all the fiscal, legal, and administrative functions of the Foundation. The director should receive a salary that is fully adequate to attract an outstanding man to the post.

6. There should be an administrative office responsible to the director to handle in one place the fiscal, legal, personnel, and other similar administrative functions necessary to the accomplishment of the purposes of the Foundation.

7. With the exception of the director, the division members, and one executive officer appointed by the director to administer the affairs of each division, all employees of the Foundation should be appointed under Civil Service regulations.

III. Organization

1. In order to accomplish the purposes of the Foundation the Members should establish several professional Divisions to be responsible

to the Members. At the outset these Divisions should be:

a. Division of Medical Research.—The function of this Division should be to support medical research.

b. Division of Natural Sciences.—The function of this Division should be to support research in the physical and natural sciences.

c. Division of National Defense.—It should be the function of this Division to support long-range scientific research on military matters.

d. Division of Scientific Personnel and Education.—It should be the function of this Division to support and to supervise the grant of scholarships and fellowships in science.

e. Division of Publications and Scientific Collaboration.—This Division should be charged with encouraging the publication of scientific knowledge and promoting international exchange of scientific information.

2. Each Division of the Foundation should be made up of at least five members, appointed by the Members of the Foundation. In making such appointments the Members should request and consider recommendations from the National Academy of Sciences which should be asked to establish a new National Research Foundation nominating committee in order to bring together the recommendations of scientists in all organizations. The chairman of each Division should be appointed by the Members of the Foundation.

3. The division Members should be appointed for such terms as the Members of the Foundation may determine, and may be reappointed at the discretion of the Members. They should receive their expenses and compensation for their services at a per diem rate of, say, $50 while engaged on business of the Foundation, but no division member should receive more than, say, $10,000 compensation per year.

4. Membership of the Division of National Defense should include, in addition to, say, five civilian members, one representative designated by the Secretary of War, and one representative of the Secretary of the Navy, who should serve without additional compensation for this duty.

Proposed Organization of National Research Foundation

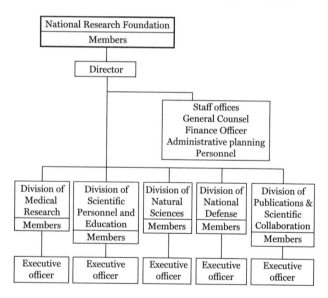

IV. FUNCTIONS

1. The Members of the Foundation should have the following functions, powers, and duties:

a. To formulate over-all policies of the Foundation.

b. To establish and maintain such offices within the United States, its territories and possessions, as they may deem necessary.

c. To meet and function at any place within the United States, its territories and possessions.

d. To obtain and utilize the services of other Government agencies to the extent that such agencies are prepared to render such services.

e. To adopt, promulgate, amend, and rescind rules and regulations to carry out the provisions of the legislation and the policies and practices of the Foundation.

f. To review and balance the financial requirements of the several Divisions and to propose to the President the annual estimate for the funds required by each Division. Appropriations should be earmarked for the purposes of specific Divisions, but the Foundation should be left discretion with respect to the expenditure of each Division's funds.

g. To make contracts or grants for the conduct of research by negotiation without advertising for bids.

And with the advice of the National Research Foundation Divisions concerned—

h. To create such advisory and cooperating agencies and councils, state, regional, or national, as in their judgment will aid in effectuating the purposes of the legislation, and to pay the expenses thereof.

i. To enter into contracts with or make grants to educational and nonprofit research institutions for support of scientific research.

j. To initiate and finance in appropriate agencies, institutions, or organizations, research on problems related to the national defense.

k. To initiate and finance in appropriate organizations research projects for which existing facilities are unavailable or inadequate.

l. To establish scholarships and fellowships in the natural sciences including biology and medicine.

m. To promote the dissemination of scientific and technical information and to further its international exchange.

n. To support international cooperation in science by providing financial aid for international meetings, associations of scientific societies, and scientific research programs organized on an international basis.

o. To devise and promote the use of methods of improving the transition between research and its practical application in industry.

2. The Divisions should be responsible to the Members of the Foundation for—

a. Formulation of programs and policy within the scope of the particular Divisions.

b. Recommendations regarding the allocation of research programs among research organizations.

c. Recommendation of appropriate arrangements between the Foundation and the organizations selected to carry on the program.

d. Recommendation of arrangements with State and local authorities in regard to cooperation in a program of science scholarships and fellowships.

e. Periodic review of the quality of research being conducted under the auspices of the particular Division and revision of the program of support of research.

f. Presentation of budgets of financial needs for the work of the Division.

g. Maintaining liaison with other scientific research agencies, both governmental and private, concerned with the work of the Division.

V. PATENT POLICY

The success of the National Research Foundation in promoting scientific research in this country will depend to a very large degree upon the cooperation of organizations outside the Government. In making contracts with or grants to such organizations the Foundation should protect the public interest adequately and at the same time leave the cooperating organization with adequate freedom and incentive to conduct scientific research. The public interest will normally be adequately protected if the Government receives a royalty-free license for governmental purposes under any patents resulting from work financed by the Foundation. There should be no obligation on the research institution to patent discoveries made as a result of support from the Foundation. There should certainly not be any absolute requirement that all rights in such discoveries be assigned to the Government, but it should be left to the discretion of the director and the interested Division whether in special cases the public interest requires such an assignment. Legislation on this point should leave to the Members of the Foundation discretion as to its patent policy in order that patent arrangements may be adjusted as circumstances and the public interest require.

VI. Special Authority

In order to insure that men of great competence and experience may be designated as Members of the Foundation and as members of the several professional Divisions, the legislation creating the Foundation should contain specific authorization so that the Members of the Foundation and the Members of the Divisions may also engage in private and gainful employment, notwithstanding the provisions of any other laws: provided, however, that no compensation for such employment is received in any form from any profit-making institution which receives funds under contract, or otherwise, from the Division or Divisions of the Foundation with which the individual is concerned. In normal times, in view of the restrictive statutory prohibitions against dual interests on the part of Government officials, it would be virtually impossible to persuade persons having private employment of any kind to serve the Government in an official capacity. In order, however, to secure the part-time services of the most competent men as Members of the Foundation and the Divisions, these stringent prohibitions should be relaxed to the extent indicated.

Since research is unlike the procurement of standardized items, which are susceptible to competitive bidding on fixed specifications, the legislation creating the National Research Foundation should free the Foundation from the obligation to place its contracts for research through advertising for bids. This is particularly so since the measure of a successful research contract lies not in the dollar cost but in the qualitative and quantitative contribution which is made to our knowledge. The extent of this contribution in turn depends on the creative spirit and talent which can be brought to bear within a research laboratory. The National Research Foundation must, therefore, be free to place its research contracts or grants not only with those institutions which have a demonstrated research capacity but also with other institutions whose latent talent or creative atmosphere affords promise of research success.

As in the case of the research sponsored during the war by the Office of Scientific Research and Development, the research sponsored by the National Research Foundation should be conducted, in general, on an actual cost basis without profit to the institution receiving the research contract or grant.

There is one other matter which requires special mention. Since research does not fall within the category of normal commercial or procurement operations which are easily covered by the usual contractual relations, it is essential that certain statutory and regulatory fiscal requirements be waived in the case of research contractors. For example, the National Research Foundation should be authorized by legislation to make, modify, or amend contracts of all kinds with or without legal consideration, and without performance bonds. Similarly, advance payments should be allowed in the discretion of the Director of the Foundation when required. Finally, the normal vouchering requirements of the General Accounting Office with respect to detailed itemization or substantiation of vouchers submitted under cost contracts should be relaxed for research contractors. Adherence to the usual procedures in the case of research contracts will impair the efficiency of research operations and will needlessly increase the cost of the work of the Government. Without the broad authority along these lines which was contained in the First War Powers Act and its implementing Executive Orders, together with the special relaxation of vouchering requirements granted by the General Accounting Office, the Office of Scientific Research and Development would have been gravely handicapped in carrying on research on military matters during this war. Colleges and universities in which research will be conducted principally under contract with the Foundation are, unlike commercial institutions, not equipped to handle the detailed vouchering procedures and auditing technicalities which are required of the usual Government contractors.

VII. Budget

Studies by the several committees provide a partial basis for making an estimate of the order of magnitude of the funds required to implement the proposed program. Clearly the program should grow in a healthy manner from modest beginnings. The following very rough estimates are given for the first year of operation after the Foundation is organized and operating, and for the fifth year of operation when it is expected that the operations would have reached a fairly stable level:

Activity	Millions of dollars	
	First year	*Fifth year*
Division of Medical Research	$5.0	$20.0
Division of Natural Sciences	10.0	50.0
Division of National Defense	10.0	20.0
Division of Scientific Personnel and Education	7.0	29.0
Division of Publications & Scientific Collaboration	.5	1.0
Administration	1.0	2.5
	33.5	122.5

Action by Congress

The National Research Foundation herein proposed meets the urgent need of the days ahead. The form of the organization suggested is the result of considerable deliberation. The form is important. The very successful pattern of organization of the National Advisory Committee for Aeronautics, which has promoted basic research on problems of flight during the past thirty years, has been carefully considered in proposing the method of appointment of Members of the Foundation and in defining their responsibilities. Moreover, whatever program is established it is vitally important that it satisfy the Five Fundamentals.

The Foundation here proposed has been described only in outline. The excellent reports of the committees which studied these matters are attached as appendices. They will be of aid in furnishing detailed suggestions.

Legislation is necessary. It should be drafted with great care. Early action is imperative, however, if this nation is to meet the challenge of science and fully utilize the potentialities of science. On the wisdom with which we bring science to bear against the problems of the coming years depends in large measure our future as a nation.

Scientific Progress Is Essential

Progress in the war against disease depends upon a flow of new scientific knowledge. New products, new industries, and more jobs require continuous additions to knowledge of the laws of nature, and the application of that knowledge to practical purposes. Similarly, our defense against aggression demands new knowledge so that we can develop new and improved weapons. This essential, new knowledge can be obtained only through basic scientific research.

Science can be effective in the national welfare only as a member of a team, whether the conditions be peace or war. But without scientific progress no amount of achievement in other directions can insure our health, prosperity, and security as a nation in the modern world.

For the War Against Disease

We have taken great strides in the war against disease. The death rate for all diseases in the Army, including overseas forces, has been reduced from 14.1 per thousand in the last war to 0.6 per thousand in this war. In the last 40 years life expectancy has increased from 49 to 65 years, largely as a consequence of the reduction in the death rates of infants and children. But we are far from the goal. The annual deaths from one or two diseases far exceed the total number of American lives lost in battle during this war. A large fraction of these deaths in our civilian population cut short the useful lives of our citizens. Approximately 7,000,000 persons in the United

States are mentally ill and their care costs the public over $175,000,000 a year. Clearly much illness remains for which adequate means of prevention and cure are not yet known.

The responsibility for basic research in medicine and the underlying sciences, so essential to progress in the war against disease, falls primarily upon the medical schools and universities. Yet we find that the traditional sources of support for medical research in the medical schools and universities, largely endowment income, foundation grants, and private donations, are diminishing and there is no immediate prospect of a change in this trend. Meanwhile, the cost of medical research has been rising. If we are to maintain the progress in medicine which has marked the last 25 years, the Government should extend financial support to basic medical research in the medical schools and in universities.

For Our National Security

The bitter and dangerous battle against the U-boat was a battle of scientific techniques— and our margin of success was dangerously small. The new eyes which radar has supplied can sometimes be blinded by new scientific developments. V-2 was countered only by capture of the launching sites.

We cannot again rely on our allies to hold off the enemy while we struggle to catch up. There must be more—and more adequate—military research in peacetime. It is essential that the civilian scientists continue in peacetime some portion of those contributions to national security which they have made so effectively during the war. This can best be done through a civilian-controlled organization with close liaison with the Army and Navy, but with funds direct from Congress, and the clear power to initiate military research which will supplement and strengthen that carried on directly under the control of the Army and Navy.

And for the Public Welfare

One of our hopes is that after the war there will be full employment. To reach that goal the full creative and productive energies of the American people must be released. To create more jobs we must make new and better and cheaper products. We want plenty of new, vigorous enterprises. But new products and processes are not born full-grown. They are founded on new principles and new conceptions which in turn result from basic scientific research. Basic scientific research is scientific capital. Moreover, we cannot any longer depend upon Europe as a major source of this scientific capital. Clearly, more and better scientific research is one essential to the achievement of our goal of full employment.

How do we increase this scientific capital? First, we must have plenty of men and women trained in science, for upon them depends both the creation of new knowledge and its application to practical purposes. Second, we must strengthen the centers of basic research which are principally the colleges, universities, and research institutes. These institutions provide the environment which is most conducive to the creation of new scientific knowledge and least under pressure for immediate, tangible results. With some notable exceptions, most research in industry and Government involves application of existing scientific knowledge to practical problems. It is only the colleges, universities, and a few research institutes that devote most of their research efforts to expanding the frontiers of knowledge.

Expenditures for scientific research by industry and Government increased from $140,000,000 in 1930 to $309,000,000 in 1940. Those for the colleges and universities increased from $20,000,000 to $31,000,000, while those for the research institutes declined from $5,200,000 to $4,500,000 during the same period. If the colleges, universities, and research institutes are to meet the rapidly increasing demands of industry and Government for new scientific knowledge, their basic research should be strengthened by use of public funds.

For science to serve as a powerful factor in our national welfare, applied research both in Government and in industry must be vigorous. To improve the quality of scientific research within the Government, steps should be taken to modify the procedures for recruiting, classifying, and compensating scientific personnel in order to reduce the present handicap of governmental scientific bureaus in competing with industry and the universities for top-grade scientific talent. To provide coordination of the common scientific activities of these governmental agencies as to policies and budgets, a permanent Science Advisory Board should be created to advise the executive and legislative branches of Government on these matters.

The most important ways in which the Government can promote industrial research are to increase the flow of new scientific knowledge through support of basic research, and to aid in the development of scientific talent. In addition, the Government should provide suitable incentives to industry to conduct research, (a) by clarification of present uncertainties in the Internal Revenue Code in regard to the deductibility of research and development expenditures as current charges against net income, and (b) by strengthening the patent system so as to eliminate uncertainties which now bear heavily on small industries and so as to prevent abuses which reflect discredit upon a basically sound system. In addition, ways should be found to cause the benefits of basic research to reach industries which do not now utilize new scientific knowledge.

We Must Renew Our Scientific Talent

The responsibility for the creation of new scientific knowledge—and for most of its application—rests on that small body of men and women who understand the fundamental laws of nature and are skilled in the techniques of scientific research. We shall have rapid or slow advance on any scientific frontier depending on the number of highly qualified and trained scientists exploring it.

The deficit of science and technology students who, but for the war, would have received bachelor's degrees is about 150,000. It is estimated that the deficit of those obtaining advanced degrees in these fields will amount in 1955 to about 17,000—for it takes at least 6 years from college entry to achieve a doctor's degree or its equivalent in science or engineering. The real ceiling on our productivity of new scientific knowledge and its application in the war against disease, and the development of new products and new industries, is the number of trained scientists available.

The training of a scientist is a long and expensive process. Studies clearly show that there are talented individuals in every part of the population, but with few exceptions, those without the means of buying higher education go without it. If ability, and not the circumstance of family fortune, determines who shall receive higher education in science, then we shall be assured of constantly improving quality at every level of scientific activity. The Government should provide a reasonable number of undergraduate scholarships and graduate fellowships in order to develop scientific talent in American youth. The plans should be designed to attract into science only that proportion of youthful talent appropriate to the needs of science in relation to the other needs of the nation for high abilities.

Including Those in Uniform

The most immediate prospect of making up the deficit in scientific personnel is to develop the scientific talent in the generation now in uniform. Even if we should start now to train the current crop of high-school graduates none would complete graduate studies before 1951. The Armed Services should comb their records for men who, prior to or during the war, have given evidence of talent for science, and make prompt arrangements, consistent with current discharge plans, for ordering those who remain in uniform, as soon as militarily possible, to duty at institutions here and overseas where they can continue their scientific education. Moreover, the Services should see that those who study overseas have the benefit of the latest scientific information resulting from research during the war.

The Lid Must Be Lifted

While most of the war research has involved the application of existing scientific knowledge to the problems of war, rather than basic research, there has been accumulated a vast amount of information relating to the application of science to particular problems. Much of this can be used by industry. It is also needed for teaching in the colleges and universities here and in the Armed Forces Institutes overseas. Some of this information must remain secret, but most of it should be made public as soon as there is ground for belief that the enemy will not be able to turn it against us in this war. To select that portion which should be made public, to coordinate its release, and definitely to encourage its publication, a Board composed of Army, Navy, and civilian scientific members should be promptly established.

A Program for Action

The Government should accept new responsibilities for promoting the flow of new scientific knowledge and the development of scientific talent in our youth. These responsibilities are the proper concern of the Government, for they vitally affect our health, our jobs, and our national security. It is in keeping also with basic United States policy that the Government should foster the opening of new frontiers and this is the modern way to do it. For many years the Government has wisely supported research in the agricultural colleges and the benefits have been great. The time has come when such support should be extended to other fields.

The effective discharge of these new responsibilities will require the full attention of some over-all agency devoted to that purpose. There is not now in the permanent Governmental structure receiving its funds from Congress an agency adapted to supplementing the support of basic research in the colleges, universities, and research institutes, both in medicine and the natural sciences, adapted to supporting research on new weapons for both Services, or adapted to administering a program of science scholarships and fellowships.

Therefore I recommend that a new agency for these purposes be established. Such an agency should be composed of persons of broad interest and experience, having an understanding of the peculiarities of scientific research and scientific education. It should have stability of funds so that long-range programs may be undertaken. It should recognize that freedom of inquiry must be preserved and should leave internal control of policy, personnel, and the method and scope of research to the institutions in which it is carried on. It should be fully responsible to the President and through him to the Congress for its program.

Early action on these recommendations is imperative if this nation is to meet the challenge of science in the crucial years ahead. On the wisdom with which we bring science to bear in the war against disease, in the creation of new industries, and in the strengthening of our Armed Forces depends in large measure our future as a nation.

Acknowledgments

Many people have considered the history and implications for today of Bush and *Science, the Endless Frontier.* I have benefited from wise and informed discussions with some of them and am grateful for their generous sharing of time and thought. Some agree substantially with my thinking, others less so. At risk of neglecting some, I heartily thank the following people: Grayson Barber, Rodger Bybee, Joanne Carney, Robert Cook-Deegan, Angela Creager, Sandra Faber, Cary Funk, Michael Gorman, Dirk Hartog, Matt Hourihan, Maureen Kearney, Kei Koizumi, Jeff Laurenti, Jane Lubchenco, Michael Lubell, Shirley Malcom, Peter Meyers, Yuki Moore, Peter Nulty, Jim Poterba, Norma Rosado-Blake, John Sargent, Heidi Schweingruber, Sandy Shapiro,

Acknowledgments

Dan Shapiro, Toby Smith, Albert Teich, Michael Turner, Chris Volpe, Heidi Williams, Gregg Zachary, and several anonymous reviewers. I especially appreciate the patience and the excellent substantive and editorial guidance of Jessica Yao, Editor, Physical Sciences, Princeton University Press, and copyeditor Erin Hartshorn, and, especially, my wife, Margaret Lancefield; they have made the finished piece possible.

Rush D. Holt

Notes

1. Vannevar is pronounced van-EEE-var (rhymes with beaver), not VAN-eh-var. There is excellent biographical information about Bush in G. Pascal Zachary, *Endless Frontier: Vannevar Bush, Engineer of the American Century*, MIT Press, 1999; Daniel Kevles, *The Physicists: The History of a Scientific Community in Modern America*, Alfred A. Knopf, 1977; and in the brief publication of the National Academy of Sciences, *Vannevar Bush, A Biographical Memoir* by Jerome B. Wiesner, 1979. Bush's autobiographical reflections can be found in *Pieces of the Action, Arms and Free Men*, and *Science Is Not Enough*.

2. From the *New York Times*, quoted in Kevles, *The Physicists*, p. 300.

3. Albert Teich, "In Search of Evidence-Based Science Policy: From the Endless Frontier to SciSIP," *Annals of Science and Technology Policy*, vol. 2, no. 2, p. 13.

4. David M. Hart, *Forged Consensus: Science, Technology, and Economic Policy in the United States, 1921–1953*,

Princeton University Press, 1998, p. 8; Bruce L. R. Smith, *American Science Policy Since World War II*, The Brookings Institution, 1990, p. 36.

5. Bruce L. R. Smith, *American Science Policy*, p. 43.

6. *Atlantic Monthly*, July 1945.

7. A legacy of Bush is the perceived distinction among basic research, applied research, and development, although the distinction did not originate with him. Much research funded by the government is formally divided into these categories and managed separately and differently. In 1967, two decades after Bush's report, Harvard physicist Harvey Brooks, and subsequently Princeton Dean Donald Stokes, pointed out that this "obvious" one-dimensional flow from basic research to product is not what always occurs. (Donald E. Stokes, *Pasteur's Quadrant: Basic Science and Technological Innovation*, Brooking Institution Press, 1997) Research that produces applied practical results can simultaneously lead to basic understanding, which means the process can equally well go in the opposite direction from the popular model. The model fails to recognize the interlocking R&D ecosystem with strong interactions and feedback. The mind and the motivation of the researcher are rarely divided into these categories. The distinction between basic and applied research as

often used in government funding can hinder scientific progress. As expressed by Stokes, understanding and use become conflicting goals. In the practical development of products, the concept also presents problems. Entrepreneurs often speak of the "valley of death" where financial support is scarce to take a concept from discovery to commercial production. This is a manifestation of the mistaken view that innovation consists of distinct stages and a single direction of flow, rather than a complex ecosystem of interacting discovery, application, and development that needs systemic support. Whether financial support is from the government or private capital investors, a more holistic approach to research and development would be more productive. See Homer Neal, Tobin Smith, and Jennifer B. McCormick, *Beyond Sputnik U.S. Science Policy in the 21st Century*, University of Michigan Press, 2008, pp. 6–7; William B. Bonvillian, "The Problem of Political Design in Federal Innovation Organization," in Kaye Husbands Fealing, Julia Lane, John H. Marburger III, and Stephanie Shipp, eds., *The Science of Science Policy*, Stanford University Press, 2011; and Walter Isaacson, *The Innovators*, Simon and Schuster, 2014, pp. 219–21.

8. Harley Kilgore, "The Science Mobilization Bill," *Science*, vol. 98, no. 2537, August 13, 1943, pp. 151–152.

9. Vannevar Bush, "The Kilgore Bill," *Science*, vol. 98, no. 2557, December 31, 1943, pp. 571–577.

10. Kevles, *The Physicists*, p. 347.

11. Harley Kilgore, "Science and the Government," *Science*, vol. 102, no. 2660, December 21, 1945, pp. 630–638.

12. If both military and civilian research had been included in one agency it would have been a mismatch to the congressional appropriations structure, thus creating serious political problems, and furthermore, it would neither have prevented the military-civilian rivalry in the government nor the persistent dominance of military research funding that troubled Bush.

13. The strong public debate over whether to include social and behavioral sciences is discussed in Nathan Reingold, "Vannevar Bush's New Deal for Research: Or the Triumph of the Old Order," *Historial Studies in the Physical and Biological Sciences*, vol. 17, no. 2, 1987, pp. 299–344; Kevles, *The Physicists*; and Daniel J. Kevles, "The National Science Foundation and the Debate over Postwar Research Policy, 1942–1945: A Political Interpretation of *Science, the Endless Frontier*," *Isis*, vol. 68, no. 1, March 1977, pp. 4–26. The social and behavior sciences were excluded from the agency as created, and although they were added decades later, they suffered in relative funding, support, and prestige.

14. Audra J. Wolfe, *Freedom's Laboratory, the Cold War Struggle for the Soul of Science*, Johns Hopkins University Press, 2018, esp. pp. 32–33.

15. Budgetary data at http://www.aaas.org/programs/r-d -budget-and-policy.

16. *Science and Engineering Indicators*, National Science Board, 2020; and Kei Koizumi, "Evolution of Public Funding of Science in the United States from World War II to the Present," *Oxford Research Encyclopedia of Physics*, Oxford University Press, USA, March 2020, doi: 10.1093/acrefore/9780190871994.013.25.

17. Jonathan Gruber and Simon Johnson, *Jump Starting America, How Breakthrough Science Can Revive Economic Growth and the American Dream,* Hachette Public Affairs Books, 2019, p. 83, and with related discussion and data throughout; also budgetary data at www.aaas.org/programs/r-d-budget-and-policy.

18. Daniel Lee Kleinman, *Politics on the Endless Frontier*, Duke University Press, 1995, pp. 74–99 passim.

19. H. Kilgore, *Science*, vol. 102, no. 2660, December 21, 1945, p. 630.

20. See, for example, *Charting a Course for Success: America's Strategy for STEM Education*, Committee on STEM Education of the National Science and Technology Council, December 2018, The White House, Washington, DC.

https://www.whitehouse.gov/wp-content/uploads/2018
/12/STEM-Education-Strategic-Plan-2018.pdf The
national strategy presents STEM literacy as one of the
pillars of the strategic plan, along with filling the pipeline, yet most of the proposed administrative actions are
directed toward building a large, capable, and diverse
scientific workforce that can "sustain the national R&D
enterprise." The report laments that China has more
scientists and engineers than does the United States,
yet does not lament the shortcomings in the public's
embrace of science. It is telling that in the section on
science literacy the commission writes of providing science education "*even* [emphasis added] for those who
may never work in a STEM-related field." The emphasis
is on reaching and training a cohort who will go into
science and technology; the nontechnical public is a
secondary consideration.

21. Lawrence Frank, "Research After the War," *Science*, vol.
101, no. 2626, April 27, 1945, p. 433–434.

22. John L. Rudolph, *How We Teach Science, What's
Changed, and Why It Matters,* Harvard University
Press, 2019, pp. 118–119.

23. Jennifer Kavanaugh and Michael D. Rich, *Truth Decay,
An Initial Exploration of the Diminishing Role of Fact
and Analysis in American Public Life*, Rand Corporation, 2018.

24. For a discussion of public trust of scientific evidence see, for example, Naomi Oreskes, *Why Trust Science?* Princeton University Press, 2019.
25. Google Scholar word search; and Timothy Ferris, *The Science of Liberty: Democracy, Reason, and the Laws of Nature,* Harper Collins, 2010, p. 102.